Finance and Accounting for Energy Engineers

Finance and Accounting for Energy Engineers

by

S. Bobby Rauf

THE FAIRMONT PRESS, INC.

CRC Press
Taylor & Francis Group

Library of Congress Cataloging-in-Publication Data

Rauf, S. Bobby, 1956-
Finance and accounting for energy engineers / by S. Bobby Rauf.
p. cm.
Includes bibliographical references and index.
ISBN-10: 0-88173-642-2 (alk. paper)
ISBN-10: 0-88173-643-0 (electronic)
ISBN-13: 978-1-4398-5193-7 (Taylor & Francis distribution : alk. paper)
1. Engineering--Accounting. 2. Accounting. I. Title.

TA185.R38 2010
657.024'62131--dc22

2010031722

Finance and accounting for energy engineers / by S. Bobby Rauf

Published by The Fairmont Press, Inc.
700 Indian Trail
Lilburn, GA 30047
tel: 770-925-9388; fax: 770-381-9865
http://www.fairmontpress.com

Distributed by Taylor & Francis Ltd.
6000 Broken Sound Parkway NW, Suite 300
Boca Raton, FL 33487, USA
E-mail: orders@crcpress.com

Distributed by Taylor & Francis Ltd.
23-25 Blades Court
Deodar Road
London SW15 2NU, UK
E-mail: uk.tandf@thomsonpublishingservices.co.uk

Printed in the United States of America
10 9 8 7 6 5 4 3 2 1

10: 0-88173-642-2 (The Fairmont Press, Inc.)
13: 978-1-4398-5193-7 (Taylor & Francis Ltd.)

This book is dedicated to my mother Juliet Rauf and my father Agha Rauf, without whom, this wouldn't be!

Table of Contents

Preface

The purpose of this book is to provide an overview of important principles in the fields of finance and accounting. This book is written as a self study for energy engineers and managers who either lack formal training in the subjects of finance, accounting and engineering economics or simply need a means to refresh their knowledge in these subjects. This book bridges the gap between the typical business school "MBA" knowledge and its application in energy engineering, project management or manufacturing management responsibilities. Many energy engineers and technical managers feel inadequately equipped to understand certain important finance and accounting principles. Understanding of finance and accounting principles is important in interfacing and conducting business with accountants, financial analysts and members of upper management. This book is designed to familiarize energy professionals—in a relatively simple and easy to understand fashion—with decision making skills founded on financial calculations and quantitative analysis.

Each chapter in this book concludes with a list of questions or problems for self assessment and knowledge affirmation purposes. The answers to the questions are at the back of the book.

WHAT READERS CAN GAIN FROM THIS BOOK

- Skills and preparation necessary for succeeding in economics/financial portion of various certification and licensure exams, i.e. CEM, FE, PE, and many other trade certification exams.

- Better understanding of finance and accounting terms relative to the practice of engineering, engineering management, project management and manufacturing management.

- Greater confidence in interactions with finance and accounting departments or functions.

- Skills in development of effective financial justification on green-field, asset replacement, process refurbishment and energy conservation project funding requests.

- Essential skills for development of viable and effective departmental budgets.

- Enhanced comprehension of financial reports and financial analysis.

Chapter 1

Finance/Accounting Definitions and Concepts

INTRODUCTION

This chapter is dedicated to definition and illustration of finance and accounting terms and concepts; terms and concepts that serve as the foundation for business, commerce, and the energy industry. Concepts and terms that are straightforward and putative are addressed through brief and simple definitions. On the other hand, concepts that are relatively complex, somewhat technical and are energy related are explained in depth. Applications of the more abstract concepts are illustrated through simplified, yet realistic, case studies and examples. Since energy industry touches all facets of business and commerce, examples and case studies in this text reach beyond the energy spectrum and touch on a variety of business, commercial and industrial models. Finance and accounting terms that do not, commonly, correspond to the energy, utility, engineering and industrial realms are either excluded or touched on sparingly. Separate chapters are dedicated to finance and accounting topics that are profound, spawn derivative concepts, and require detailed discussion. Although many finance and accounting concepts in this chapter pertain to the entire commercial and industrial spectrum—due to main focus of this text being energy engineers—these concepts are explained and illustrated, in most cases, through energy or utility scenarios.

Annuity: Annuity is a set of, equal, periodic receipts or payments made by an investor, through a financial instrument. When annuity is received, it is as a result of lump sum or periodic payments made by an investor, through a financial instrument, into financial fund. When an annuity represents a set of equal, annual, payments made by an investor, in an effort to accumulate a certain future value or sum, it can be referred to as, simply, *annual payments.*

There are many different types of annuity products available in financial markets. Life insurance annuity, deferred annuity, immediate annuity and annuity with period certain are a few examples of annuity products.

Annual Payments: When an annuity is associated with a set of equal, annual, payments made by an investor, in an effort to accumulate a certain future value or sum, it is referred to as annual payments. Equal, periodic payments made to retire debt are also referred to as *annual payments*. See the definition for *sinking funds* for additional details on this topic.

Balloon Payment: A balloon payment is a large final payment necessary to retire a debt issue or a loan. Example: Consider a private utility firm that borrows $5 million from a commercial bank to fund a base load fuel cell electric power generating plant. The utility company's debt note or contract calls for an annual payment plan of $500,000 per year for four years. At the end of the 4-year period, the utility company promises to remit a lump sum of $4.5 million to the bank, as remaining balance of principal and interest. The $4.5 million in this case would be referred to as a balloon payment. The total interest paid in this case would be $1.5 million.

Bonds: A bond is a note or a financial instrument issued by a state, city, local government or the federal government to raise capital for their day-to-day activities or for specific projects. Such projects, typically, pertain to the development of local, state or federal infrastructure such as roads, utility systems, bridges, hospitals and higher education institutions.

A bond can also be considered as a type of loan taken out by companies. Investors, essentially, lend money to firms or institutions by purchasing the firms' or institutions' bonds. In exchange, the bond issuing entity pays an interest, or coupon, at predetermined intervals. These intervals can be semiannual or annual. Semiannual interest payments are made at six month intervals while annual interest payments are issued once a year. The debtor firm is obligated to return the principal, or the borrowed funds, on maturity date, stated on the bond.

Bonds can be issued to fund large energy projects such as hydroelectric power stations, wind energy farms and solar energy harnessing complexes. In the case of government issued bonds, the debtors are the municipalities, states or the federal government.

Interest earned on municipal bonds is generally exempt from federal tax. Interest payments are also exempt from state taxes for bonds that are bought by residents of the issuing state. Interest payments are further exempt from local tax if they are bought by residents of the locality that issued the bond. Capital gains associated with bond investments are, however, taxable. From risk point of view, municipal bonds are considered safer than corporate bonds, since a municipality is less likely to go bankrupt than a corporate entity.

There are two common types of municipal bonds: (1) general obligation municipal bonds and (2) revenue municipal bonds. The general obligation (GO) bonds are unsecured municipal bonds that are backed, simply, by the full faith and credit of the municipality. The GO bonds, generally, have maturities of at least 10 years and are paid off with funds from taxes or other fees. Revenue bonds, on the other hand, are typically issued to fund projects that will eventually create revenue directly, such as power generating plants, energy efficiency projects, dams, toll roads, stadiums, etc. The revenues collected from such projects are used to pay off the bonds.

Like the credit ratings of individuals, corporations and countries, bonds carry ratings that signify the credit worthiness or the level of security and reliability associated with specific bonds. The bond ratings range from AAA to D; with AAA being the most desirable and D being the most undesirable. The bond ratings are assigned by credit rating agencies such as Moody's, S&P (Standard and Poor's) and Fitch. Bonds with ratings below BBB/Baa are referred to as junk bonds. Junk bonds are considered to be speculative, lack security, represent a high risk of default and are subject to price volatility.

The federal government issues various bonds that differ in various respects, such as maturity period, discount terms, etc. Examples of some of the federal government bonds are Treasury Bills (or T-Bills), Series E Bonds, Series H Bonds, Series HH Bonds, Series EE.

Series EE bonds are issued at 50% of their face value and reach final maturity 30 years from issuance. The interest, in the case of EE Bonds, is added to the bond monthly basis and paid out when the holder cashes the bond. They are designed to reach face value in approximately 17 years, although an investor can hold them for up to 30 years and continue to accrue interest.

Series HH bonds are sold at a discount and mature at face value. They pay interest on semi-annual basis, as do most bonds. Issuance of Series HH bonds was terminated as of August 31, 2004. However, a significant volume of the Series HH bonds still held by investors pending maturity.

Series I bonds are issued at face value and offer variable yield, based on current inflation rate. The interest rate consists of two components: (1) The fixed rate and (2) variable rate. The fixed rate remains constant over the life of the bond. The variable rate resets every six months, from the time the bond is purchased, based on the current inflation rate. New rates take effect on May 1 and November 1 of each year.

Bonds—Par Value: A par value of a bond is the face value of the bond. In the U.S. bond markets, par value is the amount that the issuing firm, or the borrower, is obligated to pay to the bond holder at the maturity date. When a bond is worth less than its par value, it is priced at a discount; conversely when a bond is valued above its par value, the bond is priced at a premium. The relationship between par value of a bond, the interest payments and the bond price can be defined mathematically as follows:

Bond Price = Present Value of the Par Value + Present Value of Annuity of the Interest Payments, or:

$$\text{Bond Price} = P_{\text{Par Value of Bond}} + P_{\text{Annuity or Interest Payments}} \quad (1\text{-}1)$$

Or:

$$\text{Bond Price} = (F_{\text{Par Value of Bond}}) \cdot (P/F, i\%, n) + (A_{\text{Interest Payments}}) \cdot (P/A, i\%, n) \quad (1\text{-}2)$$

The format for Equation 1-2 and the definition of the components of the equation are explained in Chapter 3. At this juncture, it would suffice to note the following:

a) The factor **(P/F, i%, n)** is the financial factor for conversion of the future value, which is the face value of the bond, to its corresponding present value at a discount rate of i% and over the "n" periods to maturity of the bond. Financial factors for most operations are available in Appendix A, for varous commonly used periods and interest rates.
b) The factor **(P/A, i%, n)** is the financial factor for conversion of annuity, which in the case of bonds would be the periodic interest payments, to its corresponding present value at a discount rate of i% and over the "n" periods to maturity of the bond.

A bond is worth its par value when the price is equal to the face value.

When a bond is worth less than its par value, it is priced at a discount; conversely when a bond is valued above its par value, the bond is priced at a premium.

Bond Yield: The rate of return on a bond is called the bond yield. Bond yield is based on the sum of the interest payment, the redemption value at the bond's maturity, and the initial purchase price of the bond. As with most other financial investments, yield on a bond, in its simplest form can be determined by dividing the total return on the bond by total amount invested in the bond.

The terms ***yield*** and ***return*** are used frequently in this text to compare the performance of various financial investment instruments pertaining to energy, power or utility projects. In many references to bonds in this text, the term "portfolio" is used for instances where an investor, for the purpose of risk aversion, invests funds in collection of different bonds.

Bonds—Coupon Yield: Coupon yield is the annual interest rate established when the bond is issued.

Bonds—Current Yield: Current yield can be used to compare the interest income provided by a bond to the dividend income provided by a stock. Current yield is calculated by dividing the bond's annual coupon amount by the bond's current price. Note that this yield incorporates only the income portion of return, ignoring possible capital gains or losses. As such, current yield is more useful for investors concerned, mainly, with the current income.

Bonds—Discount Yield Versus Yield to Maturity: A **discount yield** calculation is based on the ***principal,*** or ***face value,*** and not the ***price paid*** for the bond. Furthermore, the discount yield calculation is premised on 360 days. Note that because discount yield calculation is based on the face value—serving as a larger denominator—and a 360 day, shorter year, it tends to understate the bond yield. The formula for the discount yield calculation is shown below as Equation (1-3).

Discount Yield (%) =

$$\frac{(\text{Face Value} - \text{Purchase Price})}{\text{Face Value}} \times \frac{360}{\text{Days to Maturity}} \times 100\% \qquad (1\text{-}3)$$

Yield to maturity (YTM) calculation is similar to the discount yield calculation with the exception of the fact that the former employs the price paid and not the principal. In addition, the yield to maturity is based on a 365 day year, instead of a 360-day year. Due to these two important differences, the yield to maturity provides a more accurate assessment of the bond yield as compared to the discount yield. Yield to maturity is also the most commonly used bond yield assessment method. Yield to maturity (YTM) can also be calculated using Microsoft Excel ®. In Microsoft Excel ®, the YIELDMAT function can be used to perform yield to maturity calculations. This calculation can also be performed using most financial calculators. The formula for yield to maturity calculation is shown below as Equation (1-4).

Yield to Maturity (%) =

$$\frac{\text{(Face Value – Purchase Price)}}{\text{Purchase Price}} \times \frac{365}{\text{Days to Maturity}} \times 100\% \qquad (1\text{-}4)$$

Example 1.1

An investor purchases a $10,000, face value, utility bond for $9,700. This bond has a maturity period of three (3) months. (a) What would be the discount yield on this bond? (b) Calculate and compare the yield to maturity for this bond with the discount yield.

Solution:

(a) Applying Eq. (1-1):

Discount Yield (%) =

$$\frac{(\$10{,}000 - \$9{,}700)}{\$10{,}000} \times \frac{360}{90} \times 100\%$$

Discount Yield (%) = 12%

(b) Applying Eq. 1-2:

Yield to Maturity (%) =

$$\frac{(\$10{,}000 - \$9{,}700)}{\$9{,}700} \times \frac{365}{90} \times 100\%$$

Yield to Maturity (%) = 12.54%

As expected from the discussion highlighting the difference between discount yield and the yield to maturity, the discount yield value is lower and understated, at 12%, as compared to the yield to maturity value of 12.54%.

Bond Nominal Yield: The nominal yield on a bond is the percentage (%) of interest to be paid on the bond, periodically. A bond's nominal yield is calculated by dividing the annual coupon payment by the par value of the bond. It is important to note that the current bond price is expected to be the same as its par value in the calculation of its nominal yield.

Bond's Yield to Call (YTC): A callable bond always bears some probability of being called before the maturity date. See the definition for a callable bond on the next page. Investors will realize a slightly higher yield if the called bonds are paid off at a premium. An investor in such a bond may wish to know what yield, or return, will be realized if the bond is called at a particular call date, to determine whether the prepayment risk is worthwhile. YTC can be calculated by using Excel's YIELD or IRR functions, or with a financial calculator.

Realized Yield: The realized yield of a bond should be calculated if an investor plans to hold a bond only for a certain period of time, rather than to maturity. The investor will sell the bond, at a certain price, and this projected future bond price must be estimated for the realized yield calculation. Because future prices are hard to predict, this yield measurement is only an estimation of return. This yield calculation is best performed using Excel's YIELD or IRR functions, or by using a financial calculator.

Bond Mutual Funds: As the name suggests, bond mutual funds invest in bonds and other debt securities. As such, they constitute conservative investments that aim to protect the invested principal while paying out a regular income, rather than taking on more risk in search of superior returns. If you invest in a bond fund you'll receive monthly dividends from the fund that include interest payments on the fund's underlying securities plus any capital appreciation in the prices of the portfolio's bonds. As with other types of mutual funds, bond funds have a net asset value (NAV) in form of the dollar value of one share in the fund; this is the price that investors pay or receive when they buy or sell shares in the fund.

There are three basic types of bond funds: U.S. government bond funds, municipal bond funds, and corporate bond funds. The returns of these bond funds differ according to the amount of risk inherent in each fund.

Callable Bond: A callable bond is a bond that can be paid off by the issuing entity before the stated maturity date, under a given set of conditions or circumstances. In most cases, the price will be slightly above the par value for the bond and will increase the earlier the bond is called.

Capital Expenses: Capital expenses are expense transactions associated with servicing long term capital debt. Long-term capital debt includes, but is not limited to, fixed assets, such as buildings, onsite power generation equipment, on-site waste treatment facilities, processing equipment, production equipment, HVAC equipment, effluent control equipment, etc. Although, energy EPC project related capital expense would exclude long term or fixed assets that are not utility or energy related.

Common Stock: Common stock is a security representing ownership in a corporation. Example:

Compounding: The concept of compounding represents the process of paying, receiving or accruing interest on not just the principal amount of funds but a combination of principal and interest. See Chapter 11 for additional discussion and illustration of this topic. Example: Method of compounding and computing interest on investments in a savings account.

Cost Accounting: Cost accounting is the method for determining the cost of manufactured goods. Accounting departments in firms, corporations, organizations and institutions categorize and record cost, associated with the manufacturing of products or provision of services, based on the nature of the cost and where in the process that particular cost is incurred.

Cost of Goods Sold: Cost of goods sold, sometimes abbreviated as COGS, is a financial accounting term which describes the **direct costs** attributable to the production of **goods sold** by a company. Direct costs include material cost, energy cost, utility cost and direct labor cost. For power generating firms, cost of goods sold would include the cost of fuel, i.e. coal, natural gas, and other consumable fuels. Indirect costs like marketing/advertising or R&D, research and development costs are not included in the calculation of cost of goods sold.

To illustrate the essence of COGS concept, let's consider a small utility company that generates electricity through a gas fired power generator. This firm records sales of $150,000 in a given month. It has $10,000 worth of natural gas on hand at the beginning of the month and purchases an additional $60,000 worth for production of electricity during the month. The direct labor involved, in form of the wages paid, amounts to $40,000. Fuel inventory left at the end of the billing month is valued at $10,000. The COGS and the Gross Profit for the billing month would be:

Sales: $150,000

Cost of Goods Sold:

Inventory at the beginning of the month: $ 10,000
Purchases: $60,000
Direct Labor: $40,000

Subtotal: $110,000

Less: Inventory at the end of the billing month: $10,000
Net Cost of Goods Sold: $110,000 – $10,000 = $100,000

Gross Profit on Sales = Sales—COGS
= $150,000–$100,000
= $50,000

Coupon Rate: Coupon rate is the specified interest rate or amount of interest paid by a bond. See additional discussion in the Bond section of this chapter.

Credit: In the accounting realm, the term credit means depositing of funds to an account. In that respect, the accounting use of term credit is not quite the same as the use of the word "credit" in the term credit card. Example: When funds are deposited or transferred to a project account for project execution purposes, such a transaction constitutes a credit.

Creditor: Creditor is a person, entity, organization or an institution that lends money or capital to another person, entity or an institution. Example: When an energy performance company (EPC) borrows funds from a financial institution for implementation of a project, the financial institution would be called a creditor.

Debenture: A debenture is a medium- to long-term debt instrument offered by a firm to borrow money or raise capital. In that respect, a debenture is synonymous with the term bond, with one qualification: In the US the term debenture refers to an unsecured bond. Example: Junk bonds. Such bonds tend to offer high return and little security.

Debit: In the accounting realm, the term debit means withdrawal of funds from an account. Example: When funds are withdrawn, or a purchase order is written against a project account assigned to a project, the transaction constitutes a debit. Such a transaction would be recorded as a debit in the accounting system of the firm.

Debtor: Debtor is a person, entity or an institution that borrows money or capital from another person, entity or an institution. Example: When an energy performance contracting company borrows funds from a financial institution for implementation of a project, the EPC would be called a debtor.

Discounting: Discounting can be defined as the process of determining present value of future funds, and vice and versa. The discounting process is an innate component of the time value of money. Discounting process is discussed and illustrated in Chapter 3, through the study of present value, future value, annuity and gradients.

Diluted Stock: A firm's stock is said to be diluted when a firm issues additional common shares. This additional offering of shares could be a result of secondary market offering, employees exercising stock options, or conversion of convertible bonds, preferred shares or warrants into stock. A broader definition specifies dilution as any event that reduces an investor's stock price below the initial purchase price.

Discount Rate: Discount rate is the rate of interest charged by the Federal Reserve when banks borrow from the Fed. Example: If the federal government were to embark on a government sponsored large scale photovoltaic solar energy harnessing project, funds for such a project are likely to be borrowed at current discount rate. In certain realms of discussion, especially in the commercial and financial arena, a discount rate is considered to be the interest likely to be earned through alternative investment; regardless of whether such investments are financial, commercial or industrial.

8-K Report: An 8-K Report is a document filed with the SEC, Securities and Exchange Commission, that describes a change in a firm that may affect the value of its securities. The 8-K report is defined and discussed, in detail, in Chapter 4.

EPC: EPC stands for energy performance contracting. It is an approach available to the government and the private sector for funding and implementing energy or utility efficiency enhancement measures. Chapter 11 is dedicated, exclusively, to this topic.

Equity: Equity constitutes the difference between assets and liabilities. In this context, equity is more appropriately referred to as shareholder's equity, net worth or book value. It is also considered as net worth or investment in a firm by its stockholders. Example: Ownership interest in a corporation in the form of common stock or preferred stock. In the futures trading realm, equity is the value of the securities in the account, assuming that the account is liquidated at the going market price. In the context of a brokerage account, equity is the net value of an investors account; net value being the market value of securities in the account minus any margin requirements.

EUAC: Equivalent uniform annual cost is a method that can be used to annualize costs that are encountered on non-periodic basis or have unequal lives. The EUAC method can be used to compare alternative projects that have unequal lives, involve different negative cash flows or involve costs being incurred at non-coincident times. See Chapter 11 for more detail discussion and numerical example of EUAC type analysis.

Face Amount or Face Value: Face amount or face value is the amount of debt or the principal. Face value of a bond is also the par value of the bond. In the

U.S. bond markets, face value is the amount that the issuing firm, or the borrower, is obligated to pay to the bond holder at the maturity date. Typically, corporate bonds have a face value of $1000, municipal bonds $5000, and federal bonds $10,000. In the secondary market, a bond's price varies with interest rates. If interest rates are higher than the coupon rate on a bond, the bond will be sold below face value, at a "discount."

Gradient Value or Cash Flow: A gradient value or a gradient cash flow is a cash flow that is non-standard and varies uniformly. Most gradient cash flow situations involve uniformly increasing cash flows. An example of a gradient cash flow would be a uniformly increasing maintenance expense on aging equipment. Historical records of maintenance cost data can be used, through linear regression analysis and other means to predict approximate uniform annual maintenance cost increase associated with energy producing assets. Such an annual, or periodic, uniform maintenance cost increase on the energy producing assets would be referred to as ***gradient*** cash flows. Quantitative analysis, involving gradient cash flows, are illustrated in Chapter 3 through case studies and examples depicting energy and utility scenarios.

Hedging: Simultaneous buying and selling to reduce risk is called hedging. Hedging can also be explained as risk management strategy used to limit and offset the probability of loss from fluctuations in the prices of commodities, currencies, or securities. Hedging is a means for reducing risk and exposure—during unfavorable or adverse economic conditions—without the use of insurance policies. Hedging is also used for protecting one's capital against effects of inflation. In the energy arena, hedging is conducted through futures, options, and financial swaps. Natural gas futures are traded on the New York Mercantile Exchange (NYMEX) in units of 10,000 million British thermal units (MMBtu) up to 36 months

Income–in-kind: Income-in-kind is non-monetary income such as services received instead of cash.

Indenture: Indenture is legal document that specifies terms of a bond. Indenture outlines the characteristics of the bond. It is important to understand the following terms when investing in the bond market:

- Yield to Maturity (YTM)
- Current Yield

- Nominal Yield
- Yield to Call (YTC)
- Realized Yield

These terms are explained and illustrated in this chapter, under the bonds section.

Investment Portfolio: An investment portfolio is a collection of different investments with the objective of attaining diversification. Example: If an investor wishes to invest his or her savings or retirement funds exclusively in the energy industry but still desires a measure of diversity, he or she might acquire a proportionate number of shares in each of the following segments of energy industry:

a) Oil or Fossil Fuel
b) Natural Gas
c) Wind Energy
d) Solar Energy

In such a case, the investor is said to have an investment portfolio consisting of the four businesses stated above.

Junk Bonds: Junk bonds are bonds that are considered to be speculative and lack security. Junk bonds represent a high risk of default and are subject to price volatility. Junk bonds are typically rated BBB, Baa or lower. See more details on this and other bond related topics under the bond section in this chapter.

Leverage or Financial Leverage: Leveraging is magnification of the potential return on investment. Example: Control of a large value of commodities or securities in the futures market by buying on margin and, thereby, using a relatively small investment as leverage to earn higher returns.

Limit Order: A limit order is an order placed with a broker to buy or sell at a specified price. If the entire order cannot be filled at the same time, the balance may be kept on standby for later execution; such an order would be called "resting order."

Long Call: Long call is the practice of reserving or purchasing an "option," in anticipation of a rise in stock price. See the explanation for option in this chapter.

Margin: Margin is the amount an investor must put down as "good faith" to buy securities on credit.

Margin Call: Margin call is a request by a broker for an investor to place additional funds or securities in an account as collateral against borrowed funds.

Maturity Date: Maturity date is the time at which a debt issue becomes due and the principal bust be repaid. Example: If a $1,000 bond carry's a maturity date of July 30, 2010, the bond issuing entity must remit the face value of the bond, i.e. $1,000, to the bond holder on July 30, 2010. See more discussion on this topic under the bond section in this chapter.

Measures: The term "measures" is referred to, frequently, in this text. Measures represent discrete effort, consisting of labor and material, expended to affect energy conservation, energy productivity or energy cost reduction. There is frequent, interchangeable, use of the terms "project" and "measures" throughout this text. These two terms, in the context of this book, are synonymous.

MUSH Market: The acronym "MUSH" stands for municipal, university, school and hospital market segment. This market segment includes municipal and state governments, K-12 schools, universities, colleges and hospitals. MUSH represents a major market segment in the EPC ESCO market landscape.

Non-Appropriation Clause (EPC-ESCO): The non-appropriation clause is, in effect, language that is associated with the tax-exempt lease-purchase agreements to limit the owner organization or institution's liability to the current operating budget period. So, if future funds are ***not appropriated*** for implementation of the entire plan or project, the equipment may be returned to the lender, or the ESCO, and the obligation is terminated at the end of the current operating period.

Off-peak Energy Price: Off-peak energy price is the, lower, price a consumer pays for energy usage (kWh) during scheduled off-peak hours of the day, month and year. The off-peak schedule and the corresponding energy cost rate for off-peak energy consumption are revised and published, periodically, by the utility companies; usually on annual basis. Examples of electricity contracts that reference off-peak and on-peak usage are OPT, or time of the day, and HP, hourly pricing contracts.

On-peak Energy Price: On-peak energy price is the, higher, price a consumer pays for energy usage (kWh) during scheduled on-peak hours of the day, month and year. The on-peak schedule and the corresponding energy cost rate for on-peak energy consumption are published by utility companies. Similar to off-peak energy prices, examples of electricity contracts that reference on-peak usage are OPT, or time of the day, and HP, hourly pricing contracts.

Operating Period: Operating period, typically associated with annual operating expenses, extends over twelve (12) fiscal months. Monthly utility bills, direct or indirect labor associated with manufacturing, processing and fabrication activities are some examples of operating expenses. Periods over which operating expenses are accrued, recognized and recorded in accounting journals are operating periods.

Operating Profit Margin: Operating profit margin is the percentage earned on sales before deducting interest expense and taxes.

Option: The right to buy or sell something at a specified price within a specified time period. Example: A corporation—this could be an energy company—offers stock options to its employees in recognition of superior performance. This offering consists of 100 shares of company stock options for a fixed value of $50 per share, at a point in the future when the market value reaches $80 per share. Such an option would mean that if and when the market value of this firm's stock reaches the $80 mark, the employees have the option to, simultaneously, purchase 100 share of the company stock at $50 per share and immediately sell them at $80 per share; thus making a profit of $30 per share. The overall transaction would result in a gross profit of $3,000 for the employees who execute the option.

Opportunity Cost: The monetary value of a business opportunity that is forfeited or lost is opportunity cost. Example: A natural gas distribution network is disrupted for a period of one (1) hour, and lets assume that it loses $1,000,000 worth of sales during that hour. The opportunity cost in this example would be $1,000,000. In other words, the firm lost an opportunity for realizing $1,000,000 in sales during that event.

Payback Period: Payback period is simply the amount of time it takes to recoup a certain investment. Mathematically, payback period can be defined as follows:

Payback Period = Total Investment Cost/Net Income

Or

Payback Period = 1/Simple ROI

The concept of payback is explored and demonstrated in Chapter 11.

Preferred Stock: Preferred stock is a class of stock, or a form of equity, that has prior claim to common stock on the firm's earnings and assets in case of liquidation. It is a hybrid between bonds and common stocks.

Premium: A premium is the extent to which a bond's price exceeds the face amount of the debt. Example: If a coupon rate on a bond is higher than the market interest rates, the bond may be sold at price that is higher than the face value. In such a case, the difference between the price of the bond and the face value would constitute a premium.

Principal: A principal is the amount owed or the face value of a debt. Example: If a municipality issues bonds for power generating plant expansion project at a face value of $1,000, the principal and the debt represented by such a bond would be $1,000.

Red Herring: A red herring is a preliminary prospectus or a document designed to inform potential investors or stockholders about the financial standing of the firm, prior to the "Initial Public Offering," or IPO. This document is named a "Red Herring" due to the fact that it has red lettering on the title page.

Retained Earnings: Retained earnings are earnings of a firm that have been retained instead of being distributed as dividends to stockholders.

Return: Return is the sum of income plus capital gains earned on an investment. This term is widely used and illustrated in Chapters 3 and 11, in various contexts.

Selling Short: The practice of selling short, short selling or going short applies to financial market conditions when the price or value of a stock is anticipated to fall. In such falling price conditions, some investors borrow a number of shares of the stock in question and liquidate or sell them immediately. Lets call this price $\mathbf{P_1}$. The investor secures the cash from such sale until the stock approaches its lowest value. The investor then purchases the original number of shares at the lower price $\mathbf{P_2}$ and returns them to the broker in conjunction with the borrowing and trading fees associated with the transaction. Let the fees be represented by symbol **F**. The income or revenue from the overall transaction could then be stated as follows:

$$\text{Income from Short Sale} = P_1 - P_2 - F$$

Selling Long: The practice of selling long applies to market conditions involving an anticipated rise in the price of a stock. An investor purchases shares at a lower current price in anticipation of selling them long when the price rises. If the original, lower, price is $\mathbf{P_1}$, the current, higher, price is $\mathbf{P_2}$, and the trading fees are represented by symbol F, the income from the long selling would be:

$$\text{Income from Long Sale} = P_2 - P_1 - F$$

Semiannual Compounding: Semiannual compounding is compounding based on a six month period. Example: Lets consider a situation where we need to calculate the future value F of an investment **P** that is compounded on semiannual basis at a semiannual interest rate of **i%**, for **n** years. The mathematical formula for determining the future value for this case would be:

$$F = P\ (F/P,\ i\%,\ 2.n)$$

Service Units: The number of times a specific service is provided, as a product. Example: In the energy arena, these units could be Btus, DTs, Cu-ft, MCF, kWhs, kWs.

Simple Interest: If interest is accrued only on the original, principal, amount of investment or loan, it is said to be simple interest. Most interest considerations in the industrial, commercial or energy arena is compounded interest and not simple interest.

Sinking Fund: A sinking fund is a fund or account into which annual, or periodic, payments or deposits are made in an effort to accumulate a certain future value or sum, *F*. The payments credited or made into a sinking fund are annuities. These payments would be referred to as ***annual payments*** if they are equal and if they are transacted at 1-year intervals. Example: If a municipality, state or the federal government opts to raise the needed capital for a renewable energy project, i.e. a wind turbine farm, it could do so through issuance of financial instrument such as a note or a bond. In this case, the bond would constitute a sinking fund and the interest payments made to retire the debt—if made on annual basis—would be ***annual payments.***

Sinking Fund Factor: A sinking fund is simply a financial factor (A/F, i%, n) used to derive an annual payment, utilizing the formula:

$$A = F\,(A/F,\ i\,\%,\ n)$$

Spot Price: A spot price is the current price of a commodity.

Stockholders' Equity: Stockholders' equity is stockholders' investment in a firm; the sum of stock, paid-in capital and retained earnings.

Stop Loss Order: A purchase or sell order designed to limit an investor's loss on a position in a security.

10-K Report: A 10-K Report is a required annual report filed with the SEC by publicly held firms. See Chapter 4 for detailed discussion and illustration on 10-K Reports.

10-Q Report: A 10-Q Report is a required quarterly report filed with the SEC by publicly held firms See Chapter 4 for detailed discussion and illustration on this topic.

Underwriting: Underwriting is the guaranteeing of the sale of a new issue of securities, or stocks. Underwriting is the process most firms undertake in the process of "going public." When a firm goes public and sells its stock for the first time, the process is referred to as an IPO, or Initial Public Offering. Firms typically contract underwriting financial institution, or investment banks, to manage and implement their IPOs.

Working Capital: Working capital is difference between current assets and current liabilities. The concept of working capital is covered in detail in Chapter 5.

Zero Coupon Bond: A zero coupon bond is a bond that pays no interest and is, typically, sold at a discount. A zero coupon bond is also referred to as a discount rate bond or a discount bond.

ORGANIZATIONAL STRUCTURES AND COMPONENTS

Cost Center

A cost center is simply a segment of an organization or business that has the sanction to incur cost or authority to expend funds toward the production of goods or delivery of services.

Examples:

1) An electric power generating plant within an energy/utility company. An electric power generating plant, as a cost center, has the sanction to incur cost and the authority to expend funds toward the production of electrical power.

2) Molding or machining departments in an automotive engine manufacturing plant.

3) A personnel training department within a telemarketing firm.

Profit Center

A profit center is the overall business entity that not only has the authority to incur cost but has the responsibility and accountability to generate revenue and profit. A cost center could be considered a "subset" of a profit center.

Examples:

1) Consider a diversified energy company with multiple SBUs, or strategic business units. One of the SBUs could be in the electric power business. Another SBU could be in the natural gas business. This energy company could also have an SBU that is in the business of providing energy services to commercial and governmental entities. Each of these three SBUs—while a subset of the energy company—would have the authority to incur cost, purchase raw materials and services. At the same time, these SBUs also have the responsibility market and sell their individual products. In the case of the electrical SBU, the product is electrical energy; natural gas for the natural gas SBU; and products such as EPC ESCO services for the energy services SBU. Since each of the three SBUs in this example have the sanction incur cost and responsibility to generate revenue through sales, they are responsible for generating profits. Therefore, each of these three SBUs would be referred to as profit centers.

2) Pontiac is a profit center for GM; so is Chevrolet.

3) In the service realm, a machine shop or a maintenance department could be profit centers. In that, each of these two entities would provide services to internal or external customers, would incur costs, generate revenues and profits.

TYPES OF ORGANIZATIONAL STRUCTURES

Understanding of organizational structures is a critical component of finance and accounting in the energy arena. The roles of energy engineers and energy services organization are shaped, to a significant extent, by the organizational structures of their clients and service providers in the following ways:

1) An energy engineer on the client, consumer or customer side of the "fence," can conduct business with his or her counterparts on the energy supplier or "vendor" side of the fence more effectively if he or she is cognizant of the "chemistry" of the organization they are dealing with, on the other side of the fence. One important determinant in under-

standing the chemistry of an organization is knowledge of whether that organization has a relatively centralized form of management matrix or a relatively decentralized one. Such knowledge can enable the energy engineer to formulate an effective approach, identify proper protocol and develop the most suitable strategy for obtaining a bid from the vendor/ESCO that maximizes the benefits or returns on a given project.

2) If, on the other hand, you represent an energy services organization, i.e. an ESCO, formulation of a successful bid on an energy project depends, to a great degree, on how well you understand your client's organization. You portfolio of clients could include school systems, municipalities, counties, states, federal institutions, commercial organizations, consumer goods manufacturers, processing plants, industrial plants, etc. Your would need to calibrate your approach, protocol and the nature of energy measures to conform to the type of organizational culture your are encountering on the client's side.

3) Centralized and decentralized organizations are likely to use different types of accounting and financial analysis systems. Regardless of whether you are a client or a service provider, such differences could, indeed, form the basis for your overall strategy.

While organizational structures may be similar in some respect, they possess definite distinctions that reflect unique corporate cultures, business objectives, strategies, degree of autonomy at the local level and the extent of diversification.

In a centralized management matrix, organizational reporting hierarchy is such that field operations, manufacturing operations, power generation facilities or service centers are under more direct control of the corporate office and management. In centralized organizations, the local operating facilities have less autonomy and most of the significant decisions are made at the corporate offices. Also, in centralized organizations, many support services or activities are provided to the plants and field operations through central corporate offices.

Decentralized organizations tend to be diametrically opposite in their management approach as compared to the centralized organizations. The decentralized management matrices allow a great degree of autonomy and authority to local manufacturing facilities, power generating plants, local branches, local service centers, etc. Many vital business decisions, in decentralized organizations, are made at the local level. The added autonomy does, in many decentralized organizations, come at a tradeoff; in that, few or no support services are available to the local facilities through the corporate offices.

Organizational structures of businesses within the same industry tend to be similar. For instance, businesses in beverage bottling industry may be organized in a similar fashion, yet quite different from organizational structures proven to be most effective in distinctly different industry, such as the software industry. Businesses in beverage bottling industry may be significantly hierarchical, centralized and stratified. Businesses in such an industry would tend to possess multiple hierarchical tiers of management. On the other hand, in the software industry, business would tend to be staffed predominantly by professionals such as computer programmers and information technology specialists, who require less supervision. Therefore, businesses belonging to the software industry would tend to be less stratified, with fewer tiers of management and more decentralization. Such decentralized organizations allow greater autonomy to the local facilities and branch offices.

Organizational structures, centralized or decentralized, tend to evolve and transform, over time, to conform to the needs of the organization. At times, changes in organizational structures occur due, simply, to changes in leadership. Often, as organizations grow, **accretively** or **organically**, they change from being more centralized to decentralized, or vice and versa. Accretive growth is defined as growth of an organization through acquisitions. Organic growth, on the other hand consists of growth of an organization from within, through, for example, expansion of existing operations.

All organizations have some measure of hierarchy, culminating to senior executives and a chief executive officer. Even though modern businesses exemplify a stark departure from the classical models for organizations, hierarchical structures of management are still institutionalized in various industries at the highest and lowest echelons of organizations.

In our discussion and analysis of various organizational structures, we will begin with examples from the manufacturing sector to establish a broad understanding of business organizational structures or matrices. We will conclude our discussion on this subject with discussion on organizational structure of a Fortune 500 Energy corporation followed by analysis of a general, decentralized, organizational matrix in the natural gas arena. The models we will utilize for our discussion are as follows:

I. A Centralized Organizational Structure/Matrix; Manufacturing Industry—Residential and Commercial Products

II. A Decentralized Management Structure/Matrix; Manufacturing Industry—Power Electronic Products.

III. A Centralized Organizational Structure/Matrix; Energy Industry—Fortune 500 Energy Corporation.

IV. A Centralized Management Structure/Matrix, Natural Gas Industry

I. Centralized Organizational Structure/Matrix; Manufacturing Industry—Residential and Commercial Products

The organizational structure depicted in Figure 1-1 pertains to a firm that produces residential and commercial lawn care products. This firm is highly centralized. The higher management staff and the corporate officers are located in the corporate headquarters. At a glance at this organizational matrix one would infer that these four manufacturing plants are mostly ***cost centers*** with little autonomy. Most significant business decisions are made at the corporate level. Manufacturing management, frontline supervision and minimal support staff, for each of the four manufacturing plants, are located at the respective manufacturing sites. Note that all four operations report to ***one*** manufacturing director. Marketing, accounting, administrative, human resource and purchasing functions, for each of the four plants, are provided through the central or corporate offices.

II. Decentralized Organizational Structure/Matrix; Manufacturing Industry—Power Electronics Products

The model used for discussion on decentralized organizational structure/matrix in the manufacturing arena is shown in Figure 1-2. This decentralized organizational matrix pertains to a power electronics product manufacturer. While the higher management staff and the corporate officers are still located in the corporate headquarters, each plant is relatively autonomous. Each of the four plants specializes in a distinct product and is an

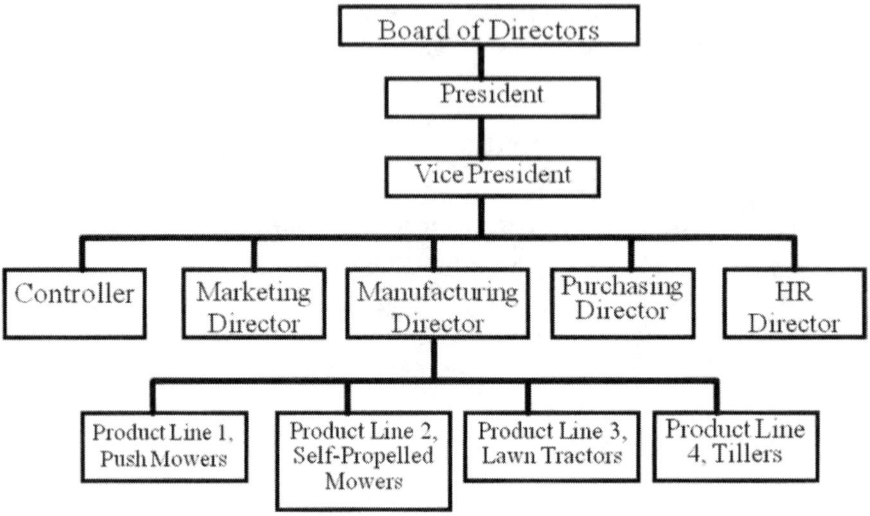

Figure 1-1. Typical Centralized Management Matrix, Manufacturing Industry—Residential and Commercial Products

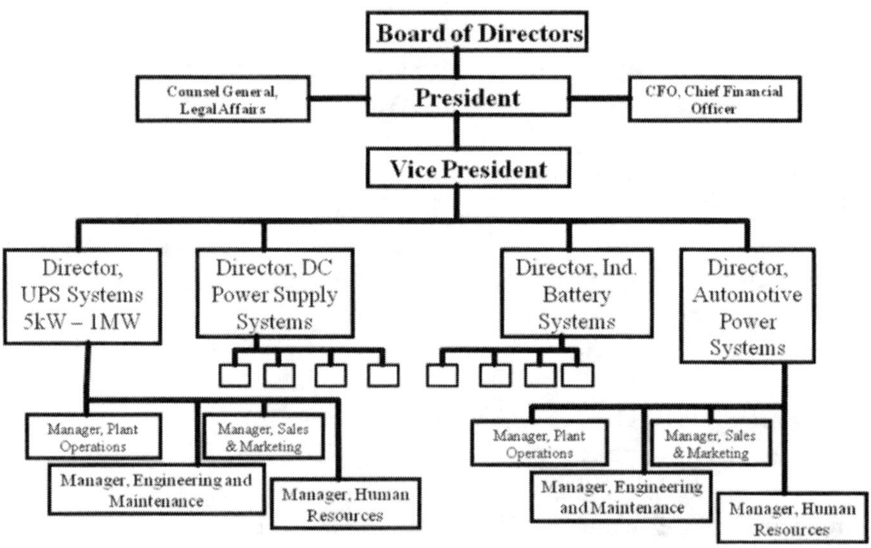

Figure 1-2. Typical Decentralized Management Matrix, Manufacturing Industry—Power Electronics

independent profit center. Each plant has the sanction for cost and the responsibility for sustaining and enhancing profitability. Each plant operation, as an independent profit center, is led by a director with a direct line reporting relationship with the corporate vice president and president. Therefore, in this decentralized organization, vital business decisions can be made at the operation or plant level. As shown in Figure 1-2, marketing, engineering, human resource—and other support functions not stated on the organization chart, i.e. accounting, information technology, purchasing, etc.—are located locally.

III. A Centralized Organizational Structure/Matrix; Energy Industry—Fortune 500 Energy Corporation.

The model used for discussion on a centralized organizational matrix in the energy industry is depicted in Figure 1-3. This centralized organizational matrix pertains to a Fortune 500 energy corporation with a national footprint. The top echelon of this centralized organization shows the CEO, chief executive officer, accountable to the board of directors. The CEO is supported by two (2) key executives whose function is vital to the sustainability and profitability of the corporation. These two executives are the CFO, chief financial officer and the counsel general, legal affairs. Important

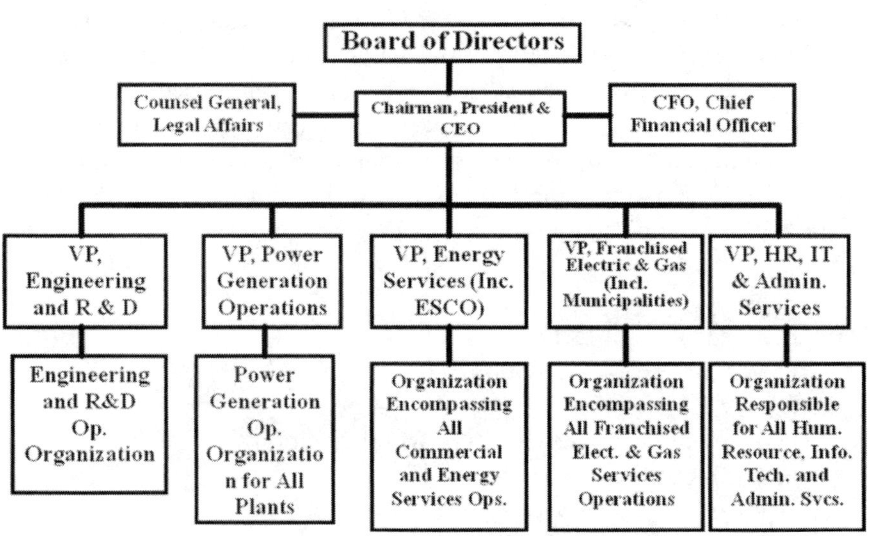

Figure 1-3. Centralized Management Matrix Example—Energy Industry

business decisions, that have ramifications across the board and have a strategic impact on all segments of the corporation, are made at the level of these three executives. This attests to the fact that this energy firm is a centralized organization.

Another noteworthy fact about this firm is that this centralization of higher level of authority is neither traditional nor longstanding. This energy firm was a relatively decentralized organization prior to the appointment of the current CEO. Upon taking the reins of the organization, a few years ago, the current CEO transformed the management matrix from a decentralized matrix to a centralized one. Drastic changes of such magnitude, with strategic corporate impact, among energy as well as non-energy firms, are not uncommon.

As we examine the organizational hierarchy in the third tier of this energy firm we notice a measure of independence and self-sufficiency. In that, each of the five functional groups—ranging from the engineering and R&D group to the HR, IT and administrative services group, is led by a vice president and supported by a fairly complete and dedicated support organization. This tier of the energy firm tends to deviate from the traditional centralized organization model. Unlike the classical or absolute model of a centralized firm and not unlike the conventional decentralized model, the support services, that lower tier operations traditionally depend on the corporate offices for, are established and made available at the lower tier operation level. Therefore, we could call this energy firm's current organizational structure a "centralized-decentralized" hybrid with centralized strategic corporate control. As you examine the lower tiers of this organization, in Figure 1-3, you notice that while the following major segments of this corporation are relatively self-sufficient and independent, they don't meet the definition of an SBU, Strategic Business Unit; in that, these major segments are not true profit centers:

1. Engineering and R&D
2. Power Generation Operations
3. Energy Services (Inc. ESCO)
4. Franchised Electric & Gas (Incl. Municipalities)
5. HR, IT & Admin. Services

Study of this energy firm's centralized organization model shows that while many firms can easily be categorized into the centralized or decentralized categories, there are examples of organizations that, more appropriately, could be considered as hybrids; albeit, leaning more on one side or the other. To be more precise, therefore, this Fortune 500 firm's centralized organization could be categorized as a central leaning centralized-decentralized hybrid organization.

IV. Centralized Management Matrix; Natural Gas Industry

The model used for this centralized management matrix in the natural gas industry is shown in Figure 1-4. Note that the top echelon of this centralized organization shows the president/CEO, accountable to the board of directors. Similar to the previous centralized organizational example of the Fortune 500 energy company, the president/CEO in this organization is supported by two (2) key executives whose function is vital to the sustainability and profitability of the corporation. These two executives, as before, are the CFO, chief financial officer and the counsel general, legal affairs. Important business decisions, that have ramifications across the board and have a strategic impact at the corporate level, are made at the level of these three executives and the board.

As we examine the organizational structure of this firm in the natural gas industry, we notice that, unlike the previous energy organization example, the sales and marketing function for the entire company is centralized

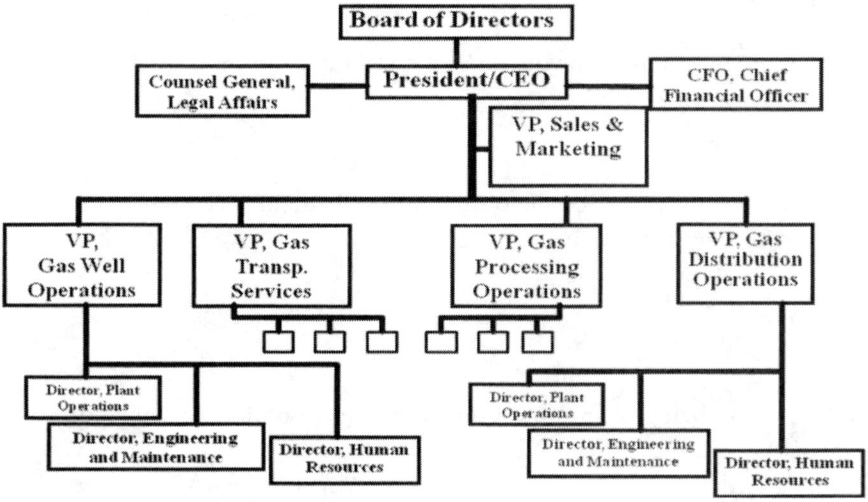

Figure 1-4. Centralized Management Matrix—Natural Gas Industry

at the corporate level, headed by VP, sales and marketing. Even though this style of organization is considered centralized, support services, other than sales and marketing, are catered at the operation or business function level.

The four (4) production, refinement and distribution related operations, or business functions, are listed below:

1. Gas Well Operations
2. Gas Transp. Services
3. Gas Processing Operations
4. Gas Distribution Operations

As apparent from the mere titles and descriptions of these business functions, they could not be considered as SBUs. While these business functions are, clearly, cost centers they have no sales, marketing or revenue function. These four operations do not qualify as profit centers. The profit stewardship and accountability lies at the corporate level; thus, centralizing the control of the corporation at the top echelon. Hence, this firm's organization is, unambiguously, centralized.

SELF ASSESSMENT PROBLEMS & QUESTIONS

Chapter 1

1. A discount rate is the discount allowed on a common stock.
   A. True
   B. False

2. The Fed just lowered the prime to 2%. By definition, the discount rate would most nearly be
   A. 2%.
   B. 0%.
   C. Prime + 2%.

3. You are being paid interest on a municipal bond, on a semiannual basis, at the rate of 8%. What would the annual rate be?
   A. 8 %
   B. 16%
   C. 4%?

4. An annuity is a payment made on annual basis.
   A. True
   B. False

5. A bond has face value of $5,000, offers and interest payments of $400 annually and has a maturity period of 10 years. Discount rate is 10%. What is maximum price you should pay of one issue of this bond?

   A. $5,000
   B. $3,614
   C. $4,500
   D. $5,614

6. An investor purchases a $10,000, face value, utility bond for $9,200. This bond has a maturity period of three (3) months. (a) What would be the discount yield on this bond? (b) What is the yield to maturity for this bond?

(a) A. 35%
    B. 32%
    C. 30%
    D. 60%

(b) A. 35%
    B. 80%
    C. 30%
    D. 32%

7. A Fortune 100 US company has just built a capacitor manufacturing plant in Ukraine. This plant builds power supplies for this company, exclusively. What would the Ukrainian plant be?

   A. Contractor
   B. Cost Center
   C. Profit Center
   D. Both (A) and (C)

8. Which of the following bond/bonds carry high risk for investors?

   A. Zero coupon bond
   B. Municipal bond
   C. A "Junk" Bond
   D. Both (A) and (C)

9. A utility company generates electricity through a gas fired power generator. This firm records sales of $500,000 in a given month. It has $40,000 worth of natural gas on hand at the beginning of the month and purchases an additional $180,000 worth for production of electricity during the month. The plant pays $60,000 in production/operation wages and $50,000 in contract labor costs. Fuel inventory left at the end of the billing month is valued at $30,000. (a) Calculate the COGS for the month. (b) Determine the gross profit for the billing month.

(a) A. $100,000
    B. $200,000
    C. $80,950
    D. $60,000

(b) A. $50,000
    B. $200,000
    C. $90,950
    D. $65,800

10. In the "Types of Organizations" section of this chapter, which of the four organizations are likely to offer greater autonomy to the business or operational functions?

    A. Organization I, III and IV
    B. Organization IV
    C. Organization II
    D. Organization III

Chapter 2

Break-even Analysis

INTRODUCTION

A study of finance, accounting, and business in general, would be incomplete without due consideration of the topic of break-even analysis. When the total revenue generated by a business "breaks even" with the total cost incurred by that business, break-even point is achieved. A manufacturing firm does not achieve the state of profitability with production and sale of the first few units of product. When a firm sells its first unit, it does make profit on that unit, based on the market price and the per unit cost assigned to it. This profit is marginal profit; it simply quantifies the amount of profit this firm would make from the sale of each additional unit. A business cannot survive, grow or ensure dividends for its stockholders solely through the sale of a few units. Cost of capital to build a manufacturing facility, indirect costs, operational costs and direct costs are spread over a volume of product produced and sold. The volume of product that must be produced and sold just to satisfy the total cost is called the ***break-even volume*** or ***break-even point.***

Knowledge of the break-even point is important because it tells the firm how many units of product must be sold in order to meet the cost. When the number of units sold exceeds the break-even point, the business is profitable. During economic recessions, break-even point analysis takes on special importance. As the demand of product declines during recessions, businesses need to know the level of sales below which the overall cost of operating exceeds the revenue and the firm begins to lose money.

Even though, thus far, we have used an industrial manufacturing example to explain the break-even concept, it is just as applicable to the energy domain. For example, if an energy supplier is interested in enhancing its capacity through an energy project involving installation of a wind energy farm or a solar photovoltaic system, this energy company would need to know what the break even volume is. In this case the break-even volume would be energy in kWhs (kilowatt-hours) or MWhs (Megawatt-hours) and demand in kW (kilowatt) or MW (Megawatt). In other words, the firm would need to know the projected electricity demand, in kWh or MW that would allow the firm to break even.

BREAK-EVEN POINT, MATHEMATICAL DEFINITION AND ANALYSIS

In this section, we take definition of break-even point to the next level. We transform the concept of break even into mathematical equations. These equations, then, serve as tools for performing break-even analysis.

Profit is defined as the difference between total revenue TR and total cost TC. This can be stated in form of Equation (2-1) below:

$$\text{Profit, } Z = TR - TC \tag{2-1}$$

Total cost is composed of fixed cost c_f and total variable cost TVC. Equation (2-1) can be expanded to include these two components of total cost, resulting in Equation (2-2) below:

$$Z = TR - c_f - TVC \tag{2-2}$$

Total revenue TR can be expanded as a product of volume "v" and per unit price "p." This results in Equation (2-3) for total profit "Z."

$$Z = v.p - c_f - v.c_v \tag{2-3}$$

As mentioned earlier, at break-even point total revenue, TR, equals total cost, TC, and the profit is zero. Therefore, Equation (2-3) can be rearranged as follows:

$$0 = v.p - c_f - v.c_v$$

$$\text{or, } v = v_{BE} = c_f/(p - c_v) \tag{2-3a}$$

The variables used in Equations (2-3) and (2-3a) can be explained as follows:

- Variable "p" is per unit price. In the manufacturing arena, this price could be in \$/piece or \$/unit. In the case of the natural gas industry, the per unit price could be in \$/MMBtu, \$/MCF or \$/DT.

- Variable "v" is the volume, or number of units produced or the number of times specific service is provided. In the case of the natural gas industry, the units for volume would be MMBtu, MCF or DT. The volume, in electrical power industry, would be represented by kWh, MWh, kW or MW.

- Fixed costs (c_f), remain constant regardless of number of units produced. Fixed building or equipment lease costs, which are independent of units produced, are one example of fixed cost. Other examples of fixed cost could be fixed overhead costs, such as, R&D, research and development costs. The total fixed cost c_f, in Equation (2-2), is a sum of all fixed costs.

- Variable cost (c_v), is the per unit cost of product. This cost is composed of all costs associated with the production of one unit of product or one unit of service provided. Mathematically, a combined or all-inclusive c_v could be represented as follows:

$$c_v = c_1 + c_2 + c_3 + c_4 + c_5 + c_6 + c_7 + \ldots + c_n \qquad (2\text{-}4)$$

Where, c_1 could, for instance, be raw material cost, c_2 could be raw material preparation cost, c_3 could be fabrication cost, c_4 could be assembly cost, c_5 could be packing cost, c_6 could be inventory carrying cost, c_7 could be per unit overhead cost.

The same approach could be extrapolated to the service industry. If the electric utility service industry were used as an example within the service industry domain, then we could redefine c_1 through c_n as representing cost components such as fuel cost, power plant operating cost, power transmission or delivery cost, overhead cost, customer service cost, etc. In the case electric utility service industry, the combined c_v could be in \$/kWh or \$/kW.

- Total variable cost, TVC, is a function of volume, "v," and the total per-unit variable cost "c_v." Total variable cost can be defined mathematically as:

$$TVC = v \cdot c_v$$

- Total cost, TC, is the sum of total fixed cost plus total variable cost.

$$\text{Or, } TC = c_f + TVC$$

$$\therefore TC = c_f + v \cdot c_v$$

- Total revenue is the product of per unit price and the volume sold.

$$\therefore TR = v \cdot p$$

- Profit (Z) is the difference between total revenue, TR and total cost, TC.

$$\therefore Z = TR - TC$$

Example 2-1: Break-even Point Calculation

According to Equation (2-3a),

$$\text{Break-even Volume} = v_{BE} = c_f/(p - c_v)$$

Smith's Automotive is a brake pad manufacturer. Annual sales and cost information for this business is listed below; determine the break-even volume:

$$c_f = \$30{,}000$$

$$c_v = \$4 \text{ per pair}$$

$$p = \$20 \text{ per pair}$$

$$v_{BE} = c_f/(p - c_v)$$

$$v_{BE} = \$30{,}000/(\$20 - \$4) = 1{,}875 \text{ pairs}$$

Or, the break-even volume, vBE = 1,875 pairs

CASE STUDY 2-1: GRAPHICAL ILLUSTRATION AND COMPUTATION OF BREAK-EVEN POINT

To illustrate the concept of break-even point and a method for determination of break even point, let us consider a simple case study related to a small natural gas fired electric power generating plant. This plant caters to a small rural municipality consisting, mostly, of residential and small commercial consumers. It is assumed that, due to the small size of demand, each consumer, commercial or private, is being charged on the basis of energy usage, in kWh, with no distinction between on- or off-peak periods. It is further assumed that load factor is not an issue. We are to determine the break-even point in terms of kWh. The data pertaining to this scenario are as follows:

$$c_f = \$40,000$$

$$c_v = \$0.035/\text{kWh}$$

$$p = \$0.045/\text{kWh}$$

Applying Equation (2-3a),

Break-even Volume = $v_{BE} = c_f/(p - c_v)$

$$v_{BE} = (\$40,000)/(\$0.045/\text{kWh} - \$0.035/\text{kWh})$$

$$v_{BE} = 4,000,000 \text{ kWh}$$

This analysis is illustrated in Figure 2-1. In this illustration, the fixed cost, variable cost, total cost and total revenue functions are plotted, independently. TR = TC at the point at which the, TR, total revenue and the, TC, total cost lines intersect. This point of intersection of TR and TC is the break-even point. A line is drawn from TR/TC intersection point, perpendicular to the abscissa (or x-axis). This line intersects the x-axis at 4,000,000 kWh; which is the break-even volume.

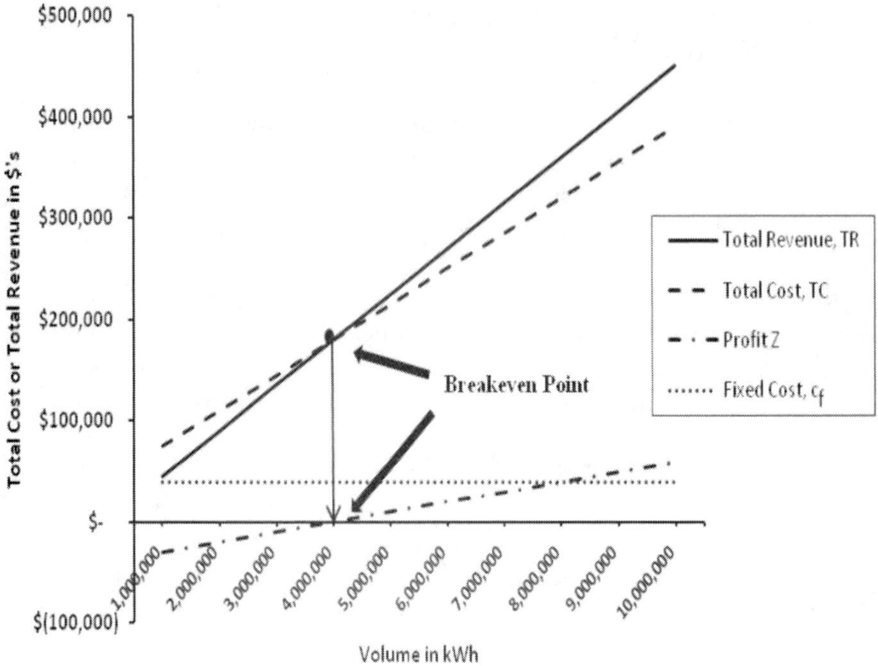

Figure 2-1. Break-even Analysis, Case Study 1

SELF ASSESSMENT PROBLEMS & QUESTIONS—CHAPTER 2

1. As per-unit price rises, the break-even volume declines.

   A. True
   B. False

2. Consider Example 1 and assume that Smith's Automotive has just identified and signed an agreement with a supplier in China that offer the raw material at 50% of the current cost. This is expected to reduce the variable cost rate, c_v, to $3.00 per pair. What would the new break-even volume, VBE, be?

   A. $1,200
   B. 3,400 pairs
   C. 1,765 pairs

3. Which of the following costs, in an industrial setting, is most certainly a fixed cost?

   A. Natural Gas Cost
   B. Wages Paid
   C. Electricity Cost
   D. Building Lease Payment

4. Profit is the same as revenue.
   A. True
   B. False

Chapter 3

Energy Engineering Economics

Time value of money based analysis of energy project investments, revenues, savings and costs. EUAC based comparison and decision making between alternative energy projects.

This chapter explains the concept of time value of money and illustrates its pertinence to the energy project-related investments, revenues from energy-related assets, costs associated with the operation of energy generating assets and the savings as a result of energy productivity improvement projects. This chapter examines how EUAC, equalized uniform annual cost method, which is premised on the fundamental concepts of time value of money, can be used as an objective and quantitative analytical tool for comparing technically equivalent energy projects or investments.

The concept of time value of money is premised on interest and interest rates. Interest is commonly understood as a fee charged by financial lending institutions, i.e. banks, to a borrower—private or commercial—for the use of a principal amount of funds. The concept of interest is a fundamental and essential entity in the absence of which the instruments, methods and usury policies that govern finance, accounting, and commerce in our society, would collapse and cease to exist.

The term discount is a term used synonymous to the term interest. Hence, the term discount rate is interest rate charged by the lending institutions to their private and commercial clients or borrowers.

TIME VALUE OF MONEY, TVM

The value of currency/money declines with passage of time as an inverse polynomial relation to the assumed interest rate (discount rate) for the period in question.

For instance, a $100 bill, 10 years in the future, will only be worth $38.55 today, based on a 10% discount rate.

METHODS AND TOOLS FOR TIME VALUE OF MONEY CALCULATIONS:

Financial Factor Tables: The financial factor method is the method used in this text for performing all time value of money calculations. The method utilizing the financial factors for time value of money calculations is demonstrated through short case studies and problems involving energy project related investments. See Appendix A for financial factor tables. The financial factors in Appendix A are premised on the formulas that govern time value of money analysis.

The terminology and acronyms used in the financial factor method, as used in this text, are streamlined with conventional financial terminology employed in energy engineering and engineering economics.

Other methods that may be used in lieu of the financial factor method are as follows:

1) Financial calculators
2) Use of spreadsheet type software, i.e. Microsoft Excel ®
3) Financial formulas

Financial Calculators: Use of financial calculators for performing calculations involving time value of money requires thorough understanding of the particular calculator's operation. Switching from one financial calculator brand to another would require additional training and understanding of brand specific features. Most multipurpose calculators do not include functions that permit gradient cash flow type calculations.

Use of Spreadsheet Type Software, i.e. Microsoft Excel ®: Similar to the use of financial calculators, use of spreadsheet type software, i.e. Microsoft

Excel ® requires thorough understanding all functions and features included on the financial function segment of the software. Formal Microsoft Excel ® software training may be recommended for some users. Maintaining the required skill level may pose some difficulty for the casual user. Some of the terminology and acronyms used in the spreadsheet type software is different from the terminology employed in energy engineering or engineering economics.

The use of spreadsheet type software does, however, offer the benefit of programmability of repetitive calculations, such as those performed in periodic cash flow based NPV, net present value, IRR, internal rate of return and payback period analysis.

Financial Formulas: The financial formula method involves the use of specific formulas for specific financial calculations.

Some of the formulas that can be used to perform time value of money analysis, pertaining to engineering economics or energy engineering economics, are stated below and explained in the subsequent section of this chapter:

$$F = P(1 - i)^n \tag{3-1}$$

$$P = \frac{F}{(1 + i)^n} \tag{3-2}$$

$$A = F \frac{i}{(1 + i)^n - 1} \tag{3-3}$$

$$F = A \frac{(1 + i)^n - 1}{i} \tag{3-4}$$

$$A = P \frac{i(1 + i)^n}{(1 + i)^n - 1} \tag{3-5}$$

$$P = A \frac{(1 + i)^n - 1}{i(1 + i)^n} \tag{3-6}$$

$$P = G\left[\frac{(1+i)^n-1}{i^2(1+i)^n} - \frac{n}{i(1+i)^n}\right] \tag{3-7}$$

$$F = G\left[\frac{(1+i)^n-1}{i^2} - \frac{n}{i}\right] \tag{3-8}$$

$$A = G\left[\frac{1}{i} \; \frac{n}{(1+i)^n-1}\right] \tag{3-9}$$

IMPORTANT TIME VALUE OF MONEY CONCEPTS— EXPLANATION AND APPLICATION OF FINANCIAL FORMULAS

Future Value, F

Value of money increases with time, based on the assumed interest, or discount, rate. Therefore, the future value of money will always be greater than the present value of money, assuming a "positive" interest or discount rate.

Figure 3-1 is a typical cash flow diagram for situation involving the calculation of a future value *F* when the present value, *P*, is known.

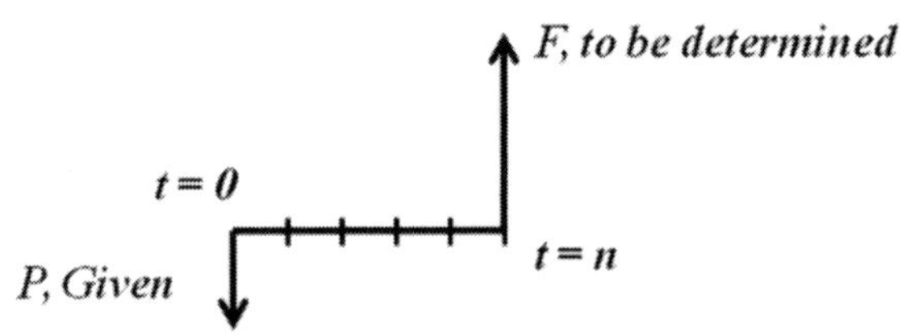

Figure 3-1.

The future value is directly proportional to the nth exponential value of the interest rate *i*, in its decimal form. Therefore, as the interest rate rises, the future value increases in magnitude. The formula used for calculating the future value is as follows:

$$F = P(1 + i)^n \tag{3-1}$$

Where,

F = Future Value
P = Present Value

i = Interest Rate, in decimal format
n = Number of Years or Compounding Periods

Note that financial factor method can be used in lieu of utilizing Equation (3-1), or the formula method. If the financial factor method were to be pursued, the mathematical operation would be as follows:

$$F = P\ (F/P,\ i\%,\ n)$$

Where, $(F/P, i\%, n)$ is the financial factor for conversion of present value to future value at interest rate of i%, in decimal format, and n number of years or pertinent compounding periods. The known, or given, present value P and the derived future value, F, would be expressed in a unit of currency, i.e. dollars. The $(F/P, i\%, n)$ financial factor, similar to all other financial factors discussed in this text, are available in Appendix A.

Present and future value analysis can be used to evaluate the relative financial worth of energy projects or investments. Such an evaluation may be necessary for deciding on whether a firm should spend funds on a project or to simply invest it in the financial market through a suitable financial instrument.

Example 3-1

At 6%, estimated, effective annual yield, how much will be accumulated if \$2,000 is invested in an energy industry stock portfolio for 10 years?

Solution:

Figure 3-2 is the cash flow diagram for Example 3-1.

$F = P\ (1+i)^n = \$2{,}000\ (1+0.06)^{10} = \$3{,}581$

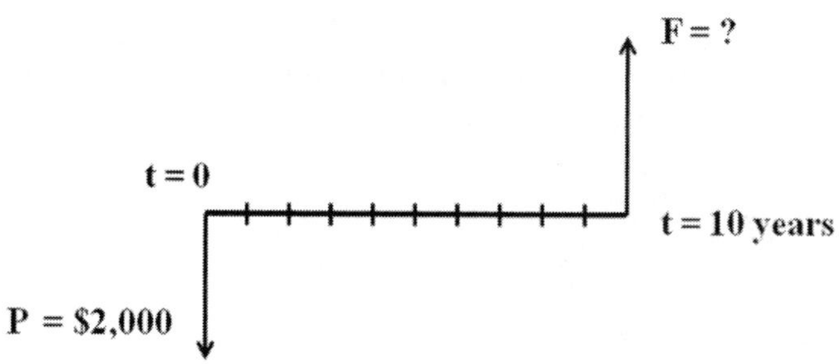

Figure 3-2

Financial Factor Method:

F = P (F/P, 6%, 10)

(F/P, 6%, 10) = 1.7908, from the financial factor tables, Appendix A

F = P (F/P, 6%, 10) = \$2,000 (1.7908) = \$3,581

Present Value, P

The present value is inversely proportional to the nth exponential value of the interest rate, *i*, in its decimal form, for a given future value *F*.

Figure 3-3 is a typical cash flow diagram for a situation involving the calculation of a present value *P* when the future value, *F*, is known. The present value *P* is represented by a downward pointing arrow. The downward direction of the arrow signifies a negative cash flow, or cash paid out, by the investor. The upward arrow represents a future value collected by the investor, after *n* periods of interest gained. Notice that the length of the present value arrow is shorter than the length of the future value arrow. This is in congruence with the expectation that the magnitude of the present value will be less than the magnitude of the future value.

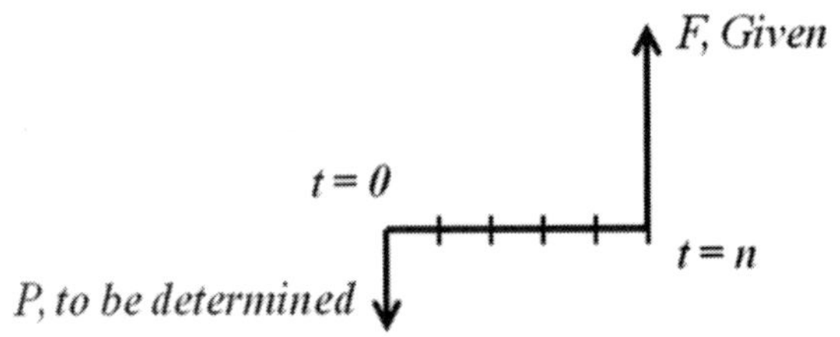

Figure 3-3

Equation (3-2) represents the formula used for calculating the present value when the future value, *F*, is known.

$$P = \frac{F}{(1 + i)^n} \tag{3-2}$$

Since the variables in Equation (3-2) are same as those in (3-1), the definitions would be the same.

Because the value of money increases with time, the present value of money is lower than the future value, with a positive discount rate. For a given future value, *F*, the present value decreases as the discount rate rises. Also, for a given future value, *F*, the present value decreases as the number of compounding periods, n, or number of years increase.

Note that financial factor method can be used in lieu of utilizing Equation (3-2), or the formula method. If the financial factor method were to be pursued, the mathematical operation would be as follows:

$$P = F\ (P/F,\ i\%,\ n)$$

Where, $(P/F,\ i\%,\ n)$ is the financial factor for conversion of future value to present value, at interest rate of $i\%$, in decimal format, and n number of years or pertinent compounding periods. The known or given, future value, F, and the derived present value, P, would be expressed in a unit of currency, i.e. dollars. The financial factor $(P/F,\ i\%,\ n)$ is retrieved from the financial factor tables in Appendix A.

Example 3-2

At 6% projected annual rate of return, how much should be invested in an energy stock portfolio today to accumulate \$3,000 in 20 years?

Solution:

$F = P\ (1+i)^n$

Or, $P = F\ (1+i) - n$

$P = \$3,000\ (1+0.6) -20 = \$935.4$

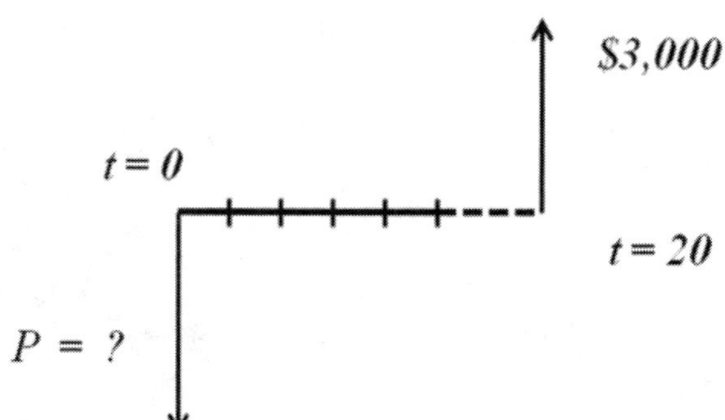

Figure 3-4

Financial Factor Method:

P = F (P/F, 6%, 20) = $3,000 (0.3118) = $935.4

Annuity, A

A simple or general form of an annuity product would be one that disburses a number of periodic payments to an investor as a result of a lump sum invested by the investor into a specific financial fund, through a specific financial instrument. This investment by the investor can also made in form of a series if payments into a specific fund. Such payments may be referred to as annuity, annual payments or, simply, payments.

Financial formulas used for calculation of annuities can be also be used in assessment of equal, periodic, payments made by an investor toward accumulation of a certain final sum or an, accumulated, future value.

Annuity calculations or annuity values can be associated with, a future value, present value or a gradient value.

$$A = F \frac{i}{(1+i)^n - 1} \tag{3-3}$$

$$F = A \frac{(1+i)^n - 1}{i} \tag{3-4}$$

$$A = P \frac{i(1+i)^n}{(1+i)^n - 1} \tag{3-5}$$

$$P = A \frac{(1+i)^n - 1}{i(1+i)^n} \tag{3-6}$$

$$A = G\left[\frac{1}{i} - \frac{n}{(1+i)^n - 1}\right] \tag{3-9}$$

Equations (3-3), (3-4), (3-5), (3-6) and (3-9) represent the mathematical relationships between annuity and future, present or gradient values associated with investments.

CONVERSION OF FUTURE VALUE TO ANNUITY

Equation (3-3) can be used to convert a known future value, *F*, to an equivalent annuity for a given interest rate *i*, for *n* number of periodic payments. This equation also represents a scenario where one is depositing a series of equal periodic payments, *A*, to build a specific sum of cash, *F*, at the end of *n* number of periods.

$$A = F \frac{i}{(1 + i)^n - 1} \qquad (3\text{-}3)$$

The cash flow diagram for this scenario is depicted in Figure 3-5. The downward pointing arrows represent annual payments, annuity or just payments. The upward arrow represents future value or the sum accumulated, and hence collected, after *n* periods or years. The scenario depicted below represents a situation where equal payments are made by the investor at the beginning of each period, or at the beginning of each year. In Figure 3-5, *F* represents the future sum or future value on the day the last payment is made. The exact interpretation of Figure 3-5 would be that annuity, or periodic payments, are being made at the beginning of each year, with the last payment being rendered on the first day of the nth year, or n^{th} period.

Note that in annuity and future value calculations, regardless of whether formulas are employed or financial factors are used, the last payment must be taken into account, notwithstanding the possibility that the future value may be liquidated at the time of the last payment, on the first day of the nth year. In other words, the financial factors and the formulas incorporate the last payment in it's principal form, with no accumulated interest.

The financial factor method can be used in lieu of the, Equation (3-

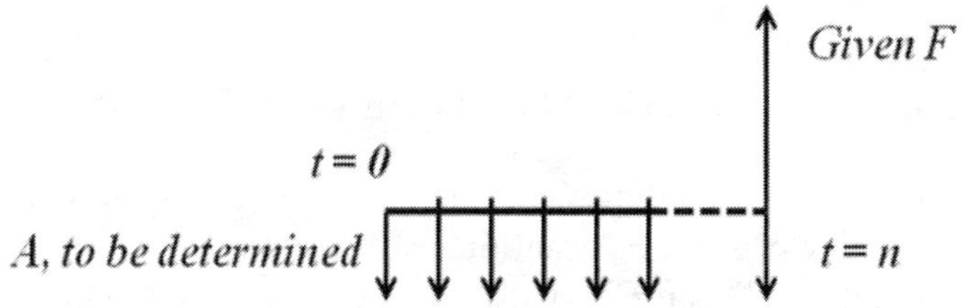

Figure 3-5

3) based, formula method. If the financial factor method were pursued, the mathematical operation would be as follows:

$$A = F\ (A/F,\ i\%,\ n)$$

Where, $(A/F,\ i\%,\ n)$ is the financial factor for conversion of future value to annuity at interest rate i, in decimal form, and n number of years or pertinent compounding periods. The known, or given, future value F and the derived annuity, or annual value, A, would be expressed in a unit of currency, i.e. dollars. The $(A/F,\ i\%,\ n)$ financial factor, similar to all other financial factors discussed in this text, are available in Appendix A.

Example 3-3

At 6% effective annual interest, how much should be deposited at the start of each year for 5 years to accumulate $30,000 at the end of 5 years?

Solution:

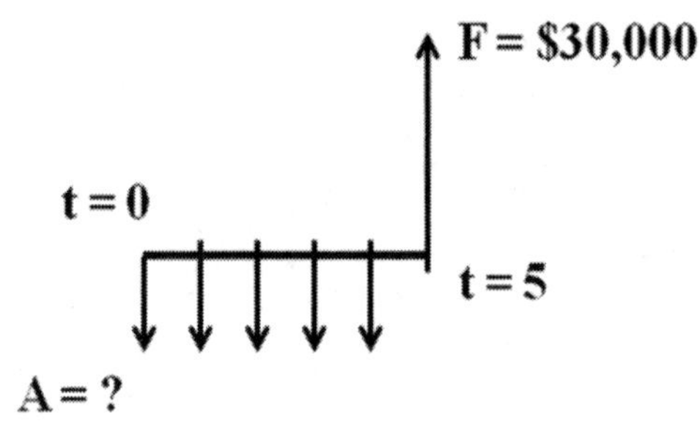

$$\begin{aligned} A &= F\ (A/F,\ 6\%,\ 5) \\ &= \$30{,}000\ (0.1774) \\ &= \$5{,}322 \end{aligned}$$

Figure 3-6

CONVERSION OF ANNUITY TO FUTURE VALUE

Equation (3-4) can be used to convert a known annuity A to a corresponding future value or sum, F, at an interest rate i, for n number of periodic annuity payments. This formula also represents situations where one is depositing a series of equal periodic payments to build a specific sum of cash at the end of n number of periods.

$$F = A \frac{(1 + i)^n - 1}{i} \tag{3-4}$$

The cash flow diagram for this scenario is depicted below in Figure 3-7. As with the cash flow diagram for Equation (3-3), the downward pointing arrows represent annual payments, annuity or just payments made. The upward arrow represents future value or the sum accumulated, and hence collected, after n periods or years.

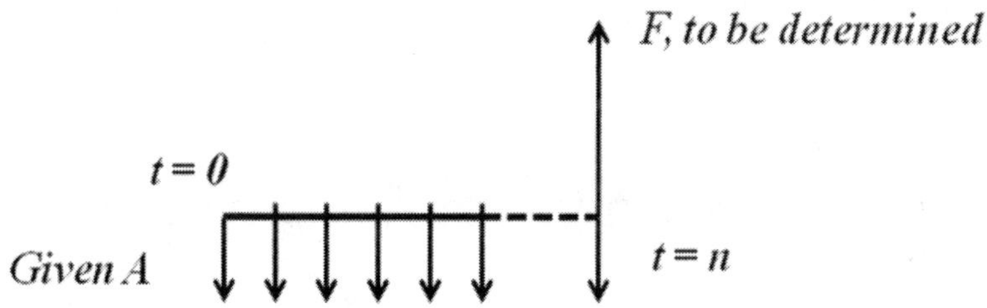

Figure 3-7

As noted in the discussion for Equation (3-3), financial factor method can be used in lieu of the formula method for determination of future value from a given or known annuity. The mathematical operation for the financial factor method would be as follows:

$$F = A\ (F/A, i\%, n)$$

Where, $(F/A, i\%, n)$ is the financial factor for conversion of annuity value to a future value or sum at interest rate i, in n number of years or compounding periods. The known, or given, annuity A and the future value F would be expressed in units of currency, i.e. dollars. The $(F/A, i\%, n)$ financial factor is available in Appendix C.

In cash flow diagrams, the conventional annuities would appear as upward pointing arrows. While the equal, periodic, payments paid toward a future lump sum value are depicted as downward pointing arrows of equal length.

Example 3-4

At 6% effective annual yield, what will be the accumulated amount at the end of 10 years if $60 is invested in a municipal bond portfolio, at the end of each year for 10 years?

Solution:

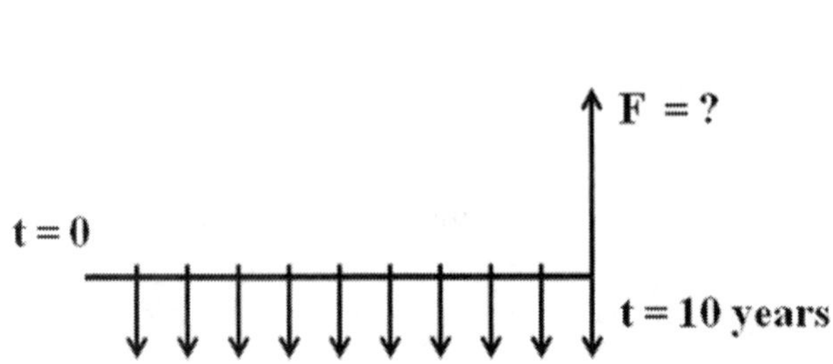

$$\begin{aligned} F &= A\ (F/A,\ 6\%,\ 10) \\ &= \$60\ (13.1808) \\ &= \$790.85 \end{aligned}$$

Figure 3-8

CONVERSION OF PRESENT VALUE TO ANNUITY

Equation (3-5) allows the conversion of a known present value, P, to a corresponding annuity at a given interest rate i, for n number of periodic payments. This equation could be utilized to determine the magnitude of annual, or equal periodic, payments that an investor can expect to receive if a certain sum of money is invested on day one of the investment period.

$$A = P\ \frac{i(1+i)^n}{(1+i)^n - 1} \tag{3-5}$$

The cash flow diagram for this scenario is depicted in Figure 3-9. The upward pointing arrows represent annual payments or annuity received. The downward arrow represents the present value equivalent, P, of n number of annuity payments. The scenario depicted below also represents a situation where equal payments, A, are received by the investor, at the end of each period, as a result of sum P invested by the investor, on the first day of the first year, or the first period. In Figure 3-9, P represents the present value or the lump sum investment which occurs on the first day of the period spanning n years.

Note that financial factor method can be used in lieu of Equation (3-5). If the financial factor method were to be pursued, the mathematical operation would be as follows:

$$A = P\ (A/P,\ i\%,\ n)$$

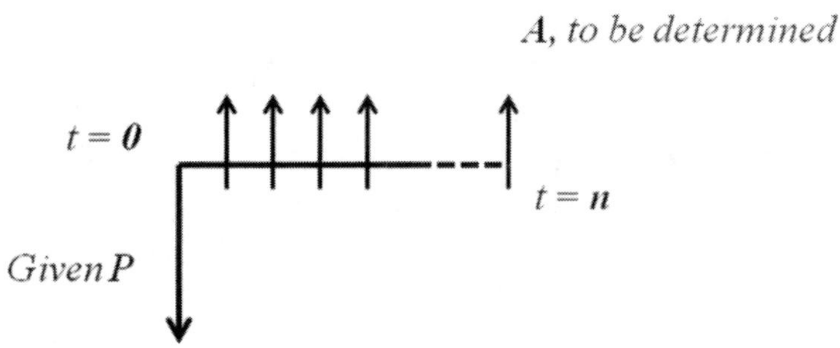

Figure 3-9

Where, $(A/P, i\%, n)$ is the financial factor for conversion of present value to annuity at interest rate i, in decimal format, and n number of years, or pertinent compounding periods. The known, or given, present value P and the derived annuity, or annual value, A, would be expressed in a unit of currency. The $(A/P, i\%, n)$ financial factor is available in Appendix A.

CONVERSION OF ANNUITY TO PRESENT VALUE

Equation (3-6) allows the conversion of a known annuity, A, to a corresponding present value, P, at a given interest rate i, for n number of periodic payments.

One practical application of this financial operation lies in the determination of the present value of a series of, equal, periodic expenditures. Fixed periodic costs, such as fixed lease payments for a mobile power generator, would be the same as annuity or periodic payments. Note that such payments constitute negative cash flows; represented by downward pointing arrows, of equal length, in the cash flow diagram.

$$P = A \frac{(1 + i)^n - 1}{i(1 + i)^n} \tag{3-6}$$

The cash flow diagram for this scenario is depicted in Figure 3-10. The long downward arrow represents the present value equivalent, P, of n number of annuity, or equal periodic payments. The present value, P, in this case could also be referred to as the cost NPV for the mobile power generator investment.

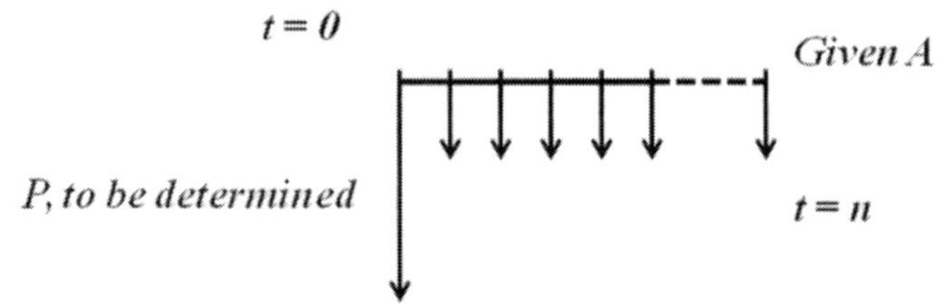

Figure 3-10

The financial factor method for derivation of the present value, from the known annuity, would be based on the following mathematical operation:

$$P = A\ (P/A,\ i\%,\ n)$$

Where, $(P/A,\ i\%,\ n)$ is the financial factor for conversion of annuity to present value, at interest rate i, and n number of years or periods. The financial factor $(P/A,\ i\%,\ n)$ is available in Appendix A, for specific periods and interests, or discount, rates.

CASE STUDY 3-1: PROJECT AND INVESTMENT DECISION BASED ON TVM ANALYSIS:

An energy firm is considering an energy project with installed, turnkey, cost of $1,000,000 and an estimated life span of 10 years. This project is expected to yield energy cost savings of $200,000 per year. The certainty of cost savings, or probability of success, is 80%. This firm needs to make a decision on whether to simply invest in a financial investment fund, Alternative I, with 70% probability of 10% return on investment or to proceed with the energy project, Alternative II. Assume a discount rate of 6%.

Solution:

A simple set of analysis can be performed to facilitate a decision between the two given alternatives. The strategy would consist of two separate analyses.

Future Value of Investment at a 10% return, Alternative I

The first analysis would involve determination of the future value of the $1,000,000 investment, over 10 years, with 10% return on investment, at 70% probability.

$$F = P\,(1 + i)^n$$

Equations representing the financial factor method for this calculation would be as follows:

$F_1 = P\,(F/P,\ 10\%,\ 10\ \text{yrs.})$

$F_1 = \$1{,}000{,}000\ (F/P,\ 10\%,\ 10\ \text{yrs.})$

$F_1 = \$1{,}000{,}000\ (2.594) = \$2{,}594{,}000$

Cash flow diagram for the first analysis is shown in Figure 3-11, below.

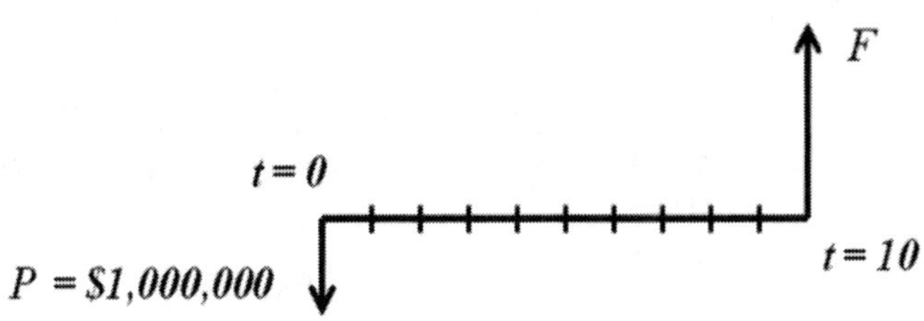

Figure 3-11

Note: The value of financial factor (*F/P*, 10%, 10 yrs.), from Appendix A, is 2.594. With 70% probability of success, the net future value, *FN*, of this alternative financial investment would be:

$F_{N1} = (\$2{,}594{,}000).\ (0.7) = \$1{,}815{,}800$

Energy Savings Per Year, for 10 Years, Alternative II

The second analysis would entail determination of the future value of the $200,000 in energy savings per year for 10 years, with a probability of success of 80% and a discount rate of 6%.

Equations representing the financial factor method for this calculation would be as follows:

$F_2 = A\ (F/A,\ 6\%,\ 10\text{yrs.})$

$F_2 = \$200{,}000\ (F/A,\ 6\%,\ 10\text{yrs.})$

$F_2 = \$200{,}000\ (13.1808) = \$2{,}636{,}160$

Cash flow diagram for the second analysis is shown in Figure 3-12.

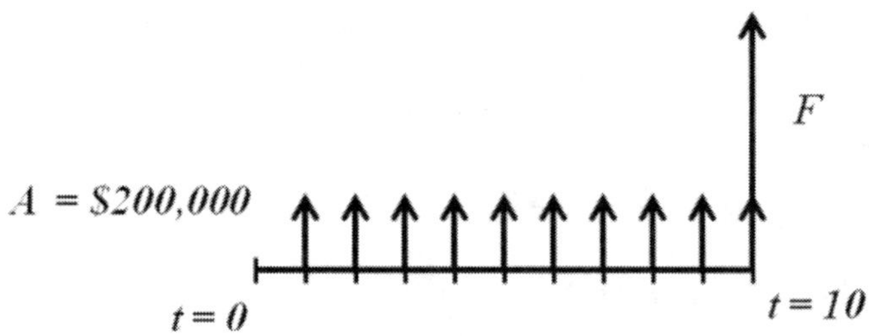

Figure 3-12

Note: The value of financial factor (*c*, 6%, 10yrs.), from Appendix A, is 13.1808. With 80% probability of success, the net future value, F_{N2}, of the second alternative would be:

$F_{N2} = (\$2{,}636{,}160) \cdot (0.8) = \$2{,}108{,}928$

Based on the two analysis conducted above, it is evident that this energy firm stands to earn, save or gain a greater future value if it proceeds forth with the energy project, versus if it just invests the $1,000,000 in a 10 year, 10% financial investment fund; a difference of ($2,108,928 –$1,815,800) or $293,128. Therefore, it is obvious that the firm would opt to engage in the project.

Note that in the analysis conducted above only the energy savings were taken into account as the sole source for positive cash flow. In reality, and energy project, such as the one illustrated in this case study, could offer other tangible benefits that might generate positive cash flow. Such tangible benefits could be in form of utility rebates, demand side management credits, lower cost of maintenance associated with newer equipment, and lower down time. Lower downtime would translate into less opportunity cost. If all of these auxiliary savings and positive cash flows were quantified and included in the calculation, the overall final or future value, with the "energy project" option would be even greater. And, the energy project would distinguish itself even more, as a more profitable alternative.

Ancillary to the Case Study 3-1

The foregoing decision making analyses were premised on comparison of the future value of the two possible alternatives. A more conventional approach to the type of scenario described in the case study would be based on the NPV, net present value, or simply, present value analysis. Since the present value of the first alternative is the current value of the available cash, no calculation is necessary, and the present value, P_1, of the first alternative would be \$1,000,000.

In order to determine the present value, P_2, of the second alternative, we must calculate the present value of the 10 equal, periodic, positive, cash flows representing the \$200,000 annual savings, over the 10-year life span of the project, at the given, or prevailing, annual discount rate of 6 %.

The present value of Alternative II can be obtained by applying Equation (3-6):

$$P_2 = A \frac{(1 + i)^n - 1}{i(1 + i)^n}$$

Or, by using the financial factor method:

$$P_2 = \text{NPV}_2 = A\ (P/A,\ 6\%,\ 10\text{yrs.})$$

According to the financial factor tables in Appendix A, the value of factor (*P*/*A*, 6%, 10yrs.) is 7.36=01.

$$\therefore\ P_2 = \text{NPV}_2 = \$200{,}000\ (7.3601)$$

$$P_2 = \text{NPV}_2 = \$1{,}472{,}020$$

With 80% probability of success for Alternative II, the net present value, P_{N2}, would be:

$$P_{N2} = \text{NPV}_{2N} = (\$1{,}472{,}020) \bullet (0.8)$$

$$P_{N2} = \text{NPV}_{2N} = \$1{,}177{,}616$$

The NPV for Alternative II is $1,177,616, versus $1,000,000 for Alternative I; a difference of $177,616. As explained in the NPV, net present value, section, when NPV is used for evaluating alternative investments, or investment options, the option or alternative that offers a greater NPV constitutes a, financially, favorable alternative. Therefore, in this case study, the NPV analysis method supports the results and decision made on the basis of the preceding future value calculation method.

CONVERSION OF GRADIENT VALUE TO PRESENT VALUE, FUTURE VALUE AND ANNUITY

As explained in Chapter 1, a gradient cash flow is cash flow that is non-standard and varies uniformly. Most gradient cash flow situations involve uniformly increasing cash flows.

Since almost all energy producing assets require various forms or maintenance, preventive maintenance, predictive maintenance and unscheduled repair, to name a few, they serve as practical and real examples of gradient cash flows. Approximate uniform annual maintenance cost increases, pertaining to energy producing assets, are examples of ***gradient*** cash flows. In present or future value analysis of annuities, gradient cash flows don't get recognized until the second year, or second period, of cash flows. The first year's cash flow represents the initial or constant cash flow. The second year's cash flow represents the sum of initial cash flow and the uniform gradient value. This process is repeated through the nth term.

Gradient cash flows or gradient values can be converted to corresponding present values, future values and annuities. Equations (3-7), (3-8) and (3-9) can be used to convert gradient cash flows to corresponding present values, future values and annuities, respectively, for a given interest rate i and n number of periodic payments.

$$P = G\left[\frac{(1+i)^n - 1}{i^2(1+i)^n} - \frac{n}{i(1+i)^n}\right] \tag{3-7}$$

$$F = G\left[\frac{(1+i)^n - 1}{i^2} - \frac{n}{i}\right] \tag{3-8}$$

$$A = G\left[\frac{1}{i} \;\; \frac{n}{(1+i)^n - 1}\right] \qquad (3\text{-}9)$$

However, it must be noted that the present value, future value and the annuity determination of gradient cash flows yield only the present, future and annuity values for the incremental, or gradient, values, and such calculations do not take into account the periodic base values. The present, future and annuity values of the base values must be assessed separately, and then added to the present, future and annuity values of the gradient values to determine the total present, future and annuity values, respectively.

Even though the gradient cash flow doesn't materialize until the second year, when gradient cash flows are being converted to present, future or annuity values, over a certain life span n, the financial factors, i.e. (*P*/G, *i*%, *n*), (F/G, *i*%, *n*) and (*A*/G, *i*%, *n*), are still selected based on the total number of periods, n. The formulas and the financial factors take into account the second year appearance of the first gradient cash flow. This rule is illustrated in Case Study 3-1.

Conversion of gradient cash flow to present value: Equation (3-7) can be used to convert a known gradient cash flow to a corresponding present value. Such an analysis could be useful in determining the present value of an investment from the cost perspective. Uniformly increasing maintenance costs in a power plant could be such gradient cash flows with a finite present value. Of course, adherence to the negative sign convention for expenses or costs, would yield a negative present value. Nevertheless, when comparing alternative investments, such as, Power Plant A versus Power Plant B, one would favor the alternative with a lower negative present value.

Figure 3-13 is a graphical depiction of a negative gradient cash flow scenario; one that might represent the uniformly increasing maintenance cost of an aging power plant, an aging manufacturing facility or an aging commercial facility.

In Figure 3-13, B represents the base value or the cost that is constant from year to year. In a power plant, any other energy facility or a manufacturing operation, B can be any cost that is constant and one that is realized or recorded at the end of the very first year. An example of B could be any type of maintenance, such as, preventative maintenance that must be conducted every year, regardless of the vintage of the equipment. As expected, being a cost transaction and a negative cash flow, B is represented as a downward

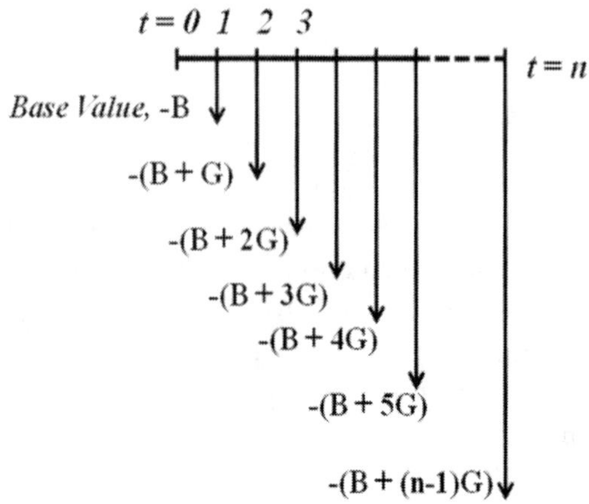

Figure 3-13

pointing negative cash flow arrow in Figure 3-13.

In Figure 3-13, G represents the incremental gradient cash flow value; a cost that increases from year to year, beginning with the second year. In a power plant example stated earlier, G could be incidental breakdown related maintenance expense; occurrence of which, typically, rises as the equipment ages. Since cost related gradient cash flows represent an expense transaction, G is shown as a downward pointing negative cash flow arrow in Figure 3-13.

As shown in Figure 3-13, total gradient cash flow formula is: B + (n–1) G, for n^{th} year. Figures 3-14 and 3-15 show a split or sectional view of Figure 3-13, in which each year's cash flow is separated into the base and the gradient cost components.

A gradient cash flow is not always a cost related, negative, cash flow. A gradient cash flow can be a revenue related positive cash flow, as well. Figure 3-16 depicts a positive gradient cash flow scenario.

CASE STUDY 3-2: POSITIVE & NEGATIVE GRADIENT CASH FLOW, DSM PROJECT:

Due to projected increase in demand and limited generating capacity in, an energy firm is soliciting DSM, demand-side management, bids from some of its largest industrial customers. One criterion for a successful bid is the NPV, net present value, of the project. NIDA Inc. is one of those large

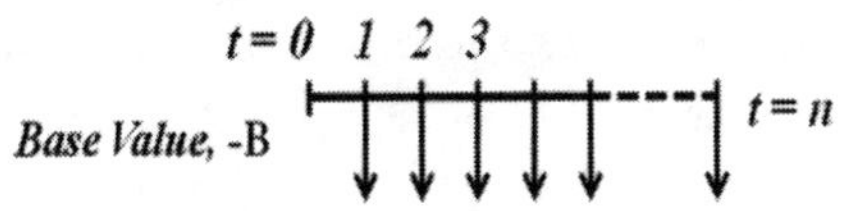

Figure 3-14

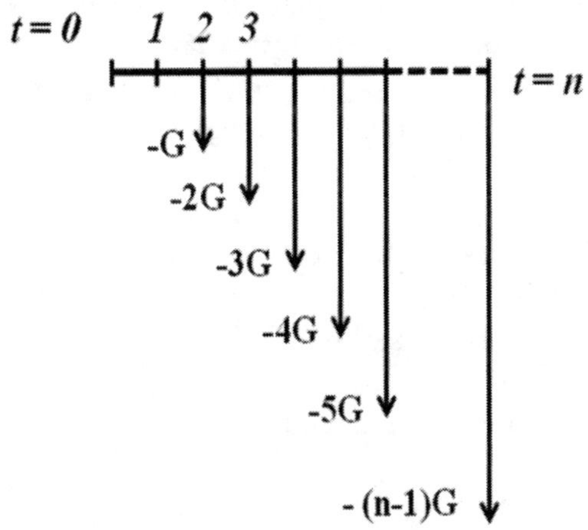

Figure 3-15

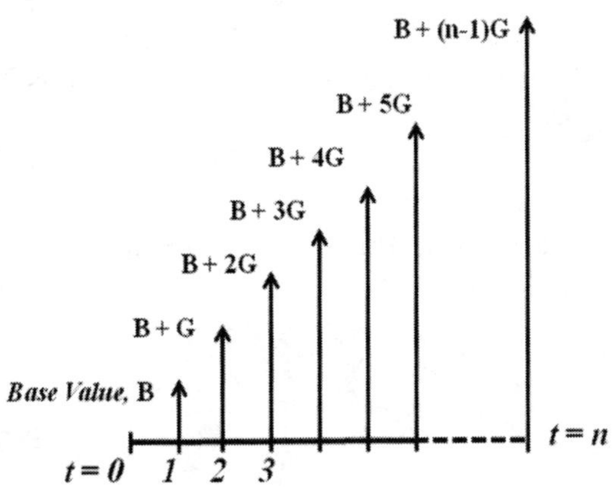

Figure 3-16

industrial customers on the energy company's grid. NIDA is considering taking advantage of this DSM bid solicitation to implement an energy conservation project. The cost, revenue and other pertinent information relative to this project are listed below:

- Initial, turnkey, cost of the project: $300,000
- Life of the project, or associated equipment: 10 years
- Expected salvage and book value of the equipment: $30,000
- DSM incentive payment, or rebate, from the energy company to NIDA: $20,000, at the end of the first year.
- The DSM terms and conditions call for DSM incentive payment is to increase by 5% each year, after the first year.
- Annual energy savings, based on the recent load profiles, are projected to average: $ 80,000
- The maintenance cost of pertinent equipment is expected to be: $6,000 at the end of the first year.
- The maintenance cost is expected to increase by 5% each year, after the first year.
- Current discount rate: 8%. Year end convention applies.

Objective: Calculate the NPV, or net present value, for the project

Solution:

In order to determine the NPV of this project, we must calculate the present values of all of the cash flows involved, negative and positive. The sum of all present values, thus calculated, would constitute the NPV for the project. In order for the project to be financially viable, the NPV must be positive.

Present Value of Cash Flow #1:

Present value of the turnkey cost of the project, registered on the last day of the fiscal year = –$300,000

$$P_{Cost} = -\$300,000$$

Present Value of Cash Flow #2:

In this case study, it is assumed that the salvage value and the book values for the project are the same. Present value of the future $30,000 salvage

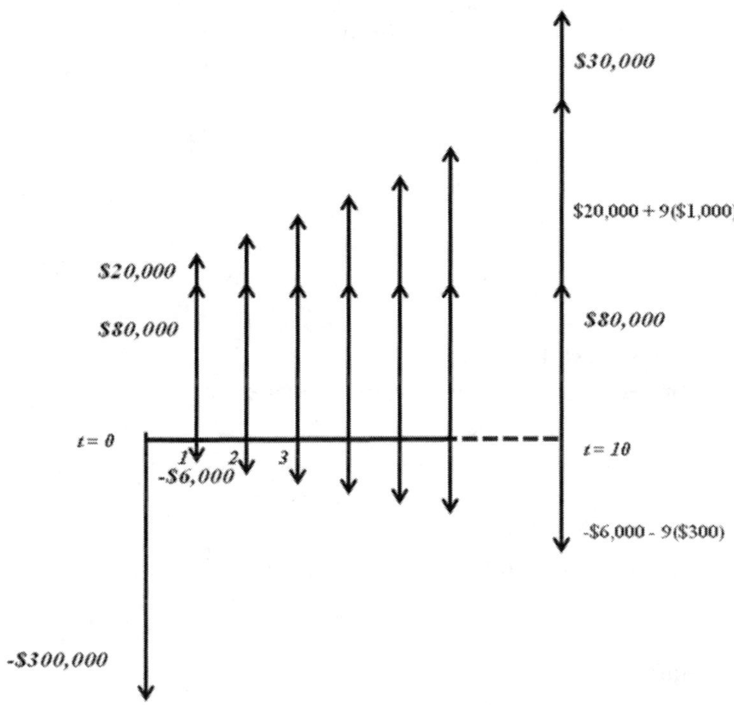

Figure 3-17

value of the project would be:

$$\begin{aligned} P_{Salvage} &= F \text{ Salvage} \cdot (P/F, i\%, n) \\ &= +\$30{,}000 \cdot (P/F, 8\%, 10) \\ &= +\$30{,}000 \cdot (0.4632), \text{ financial factor from App. A} \\ &= +\$13{,}896 \end{aligned}$$

Present Value of Cash Flow #3:

The present value of the total cash flow associated with the DSM incentive payment schedule can be calculated using the gradient cash flow method.

$$P_{DSM\ Incentive} = A_{DSM\ Incentive} \cdot (P/A, 8\%, 10) + G_{DSM\ Incentive} \cdot (P/G, 8\%, 10)$$

$$= (\$20{,}000) \cdot (P/A, 8\%, 10) + (\$20{,}000) \cdot (0.05) \cdot (P/G, 8\%, 10)$$

$$= (\$20{,}000) \cdot (6.7101) + (\$20{,}000) \cdot (0.05) \cdot (25.9768)$$

$$= \$134{,}202 + \$25{,}977$$

$$= + \$160{,}179$$

Present Value of Cash Flow #4:

The present value of the average annual energy savings of $80,000 would be:

$$P_{Energy\ Savings} = A_{Energy\ Savings} \cdot (P/A, i\%, n)$$
$$= \$80{,}000 \cdot (P/A, 8\%, 10)$$
$$= \$80{,}000 \cdot (6.7101), \text{ financial factor from App. A}$$
$$= +\$536{,}808$$

Present Value of Cash Flow #5:

The present value of the total cash flow associated with the annual maintenance cost can be calculated using the gradient cash flow method.

$$P_{Maintenance} = -A_{Maintenance} \cdot (P/A, 8\%, 10) - G_{Maintenance} \cdot (P/G, 8\%, 10)$$

$$= -(\$6{,}000) \cdot (P/A, 8\%, 10) - (\$6{,}000) \cdot (0.05) \cdot (P/G, 8\%, 10)$$

$$= -(\$6{,}000) \cdot (6.7101) - (\$6{,}000) \cdot (0.05) \cdot (25.9768)$$

$$= -\$40{,}261 - \$7{,}730$$

$$= -\$47{,}991$$

NPV for the project: The total NPV for NIDA's DSM project bid would be the sum of the present values of all cash flows associated with the project.

Total NPV for NIDA's DSM project =
– \$300,000 + \$13,896 + \$160,179 + \$536,808 – \$47,991
= \$362,892

The total NPV for NIDA's bid is a significant and positive value of \$362,892, for an initial investment of \$300,000. Of course, in a real DSM bid scenario, the energy firm might require other important parameters, i.e. the ROI, ROE and IRR to be included in the overall bid package, as well.

Conversion of gradient cash flow to future value: Equation (3-8) can be used to convert a known gradient cash flow to a corresponding future value.

$$F = G\left[\frac{(1+i)^n - 1}{i^2} - \frac{n}{i}\right] \tag{3-8}$$

Such an analysis could be useful in determining the future value of a series of revenue type positive gradient cash flows. For instance, in Case Study 3-2, application of Equation (3-8) would yield the future value of uniformly increasing DSM incentive payments by the energy company to NIDA, Inc. Conversion of a series of revenue type positive gradient cash flows to the corresponding future value can also be accomplished by utilizing the financial factor method. The general formula for this computation would be as follows:

$$F = G\ (F/G,\ i\%,\ n)$$

However, it should be noted that this computation would result in the future value of just the gradient or incremental components of each gradient cash flow. An example of incremental portion of gradient cash flows would be the periodic \$1,000 increment in the incentive payments referred to in Case Study 3-2. The future value of all of the base values, such as the annual payment of \$20,000 in Case Study 3-2, would need to be calculated separately. The future value of the gradient values and the future value of the base values would be added to obtain the total future value. In financial factor format, the total future value in a positive gradient cash flow scenario would be as follows:

$$F_{Total} = G \cdot (F/G, i\%, n) + A_{Base} \cdot (F/A_{Base}, i\%, n)$$

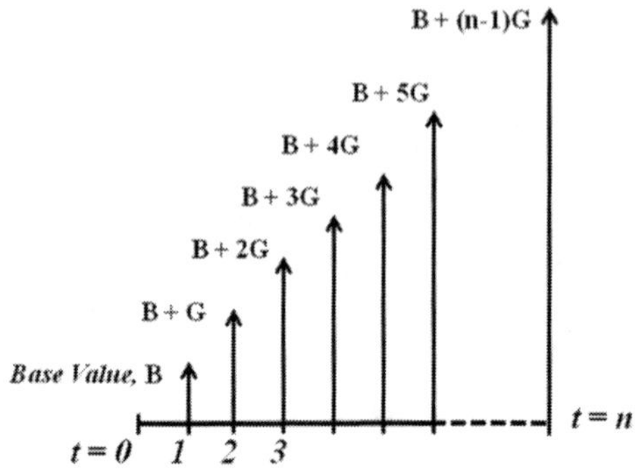

$$F_{Total} = G \,.\, (F/G, i\%, n) + A_{Base} \,.\, (F/A, i\%, n)$$

Figure 3-18

Figure 3-18 is a cash flow diagram for the combined future value, F_{Total}, while, Figures 3-19 and 3-20 represent components $A_{Base} \cdot (F/A_{Base}, i\%, n)$ and $G \cdot (F/G, i\%, n)$, respectively.

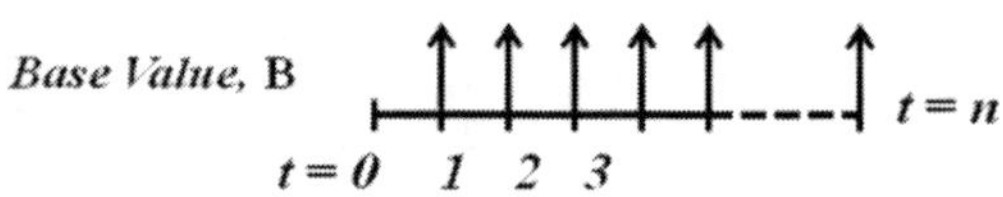

$$F_{Base} = A_{Base} \,.\, (F/A, i\%, n)$$

Figure 3-19

Conversion of gradient cash flow to annuity: Equation (3-9) can be used to convert a known gradient cash flow to a corresponding annuity.

$$A = G\left[\frac{1}{i} - \frac{n}{(1+i)^n - 1}\right] \qquad 3\text{-}9$$

Such an analysis could be useful in determining annuity, annualized value or periodic value of a series of gradient cash flows.

In Case Study 3-2, Equation (3-9) could be used to convert escalating maintenance expenses to equal uniform annualized cost equivalents, or EUAC for the maintenance expenses. In this cases study, Equation (3-9)

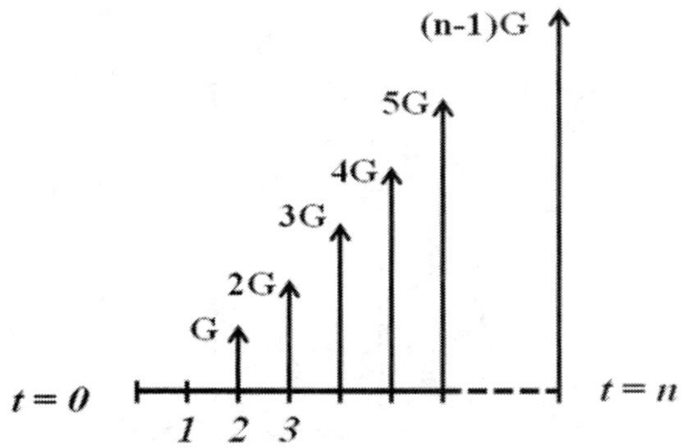

$$F_{Gradient} = G \cdot (F/G, i\%, n)$$

Figure 3-20

could also be used to convert uniformly increasing DSM incentive payments to equal uniform annualized revenue, or positive cash flow equivalents. Conversion of a series of incremental maintenance expenses or revenue type positive gradient cash flows to the corresponding annuity can also be accomplished by utilizing the financial factor method. The general formula for this computation would be as follows:

$$A = G\ (A/G, i\%, n)$$

However, it should be noted that this computation would yield annuity for just the gradient or incremental components of each gradient cash flow. The annuity equivalent for all of the base values, such as the annual payment of \$20,000 in Case Study 3-2, would be just that, i.e. \$20,000. In other words, ABase = \$20,000. The annuity equivalent of the gradient values and the base values would then be added to obtain the total annuity. In financial factor format, the total annuity in a positive gradient cash flow scenario would be as follows:

$$A_{Total} = G\ (A/G, i\%, n) + A_{Base}$$

Figure 3-21 shows the total annuity value, A_{Total}, while Figures 3-22 and 3-23 represent components A_{Base} $(F\ A_{Base}, i\%, n)$ and G $(F/G, i\%, n)$, respectively.

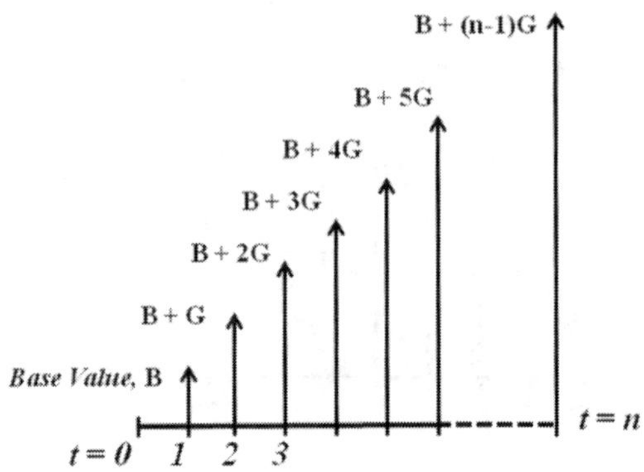

$$A_{Total} = G \cdot (A/G, i\%, n) + A_{Base}$$

Figure 3-21

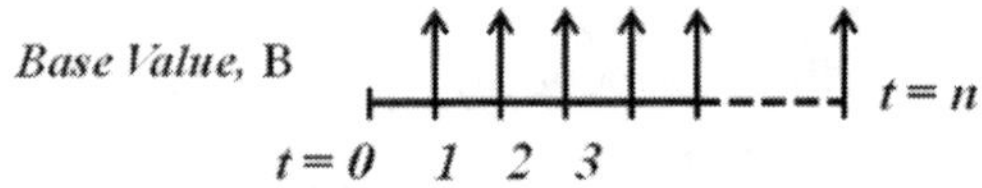

$$A_{Base} = Equal\ Annual\ Payments$$

Figure 3-22

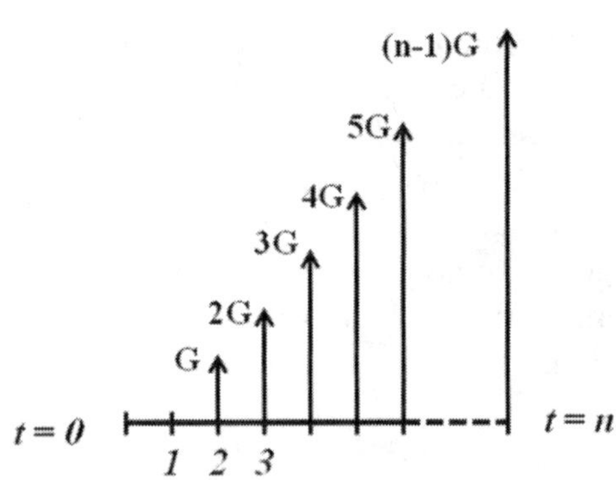

$$A_{Gradient} = G \cdot (A/G, i\%, n)$$

Figure 3-23

EUAC, A DECISION MAKING TOOL FOR ENERGY PROJECTS

As described in Chapter 1, EUAC, equivalent uniform annual cost is a method that can be used to annualize costs that are encountered on non-periodic basis or have unequal lives. The EUAC method can be used to compare alternative projects that have unequal lives, involve different negative cash flows or involve costs being incurred at non-coincident times.

As discussed in the previous section, Eq. (3-9) can be used to convert uniformly increasing maintenance expenses to equivalent uniform annual cost, or EUAC, for the maintenance expenses. As energy equipment ages, the equipment efficiency declines and the energy consumption, or energy cost, rises. EUACs for alternative projects can thus be calculated and compared to make a decision as to which project is economically superior. The method for calculating EUAC is illustrated in Example 3-5. The application of the EUAC method for deciding between alternative projects is demonstrated through Case Study 3-3.

Example 3-5

A new machine will cost $18,000 and will have a resale value of $14,000 after five years. Special tooling will cost $6000. The tooling will have a resale value of $3000 after five years. Maintenance will be $2100 per year. The effective annual interest rate is 6%. What will be the average annual cost of ownership during the next five years?

Solution:

$$EUAC = -A + F\,(A/F,\ 6\%,\ 5) - P\,(A/P,\ 6\%,\ 5)$$

$$EUAC = -\$2{,}100 + F\,(0.1774) - P\,(0.2374)$$

$$EUAC = -\$2{,}100 + \{(\$3{,}000 + \$14{,}000)\,(0.1774)\} - \{(\$18{,}000 + \$6{,}000)\,(0.2374)\}$$

$$EUAC = -\$2{,}100 + \$3{,}015 - \$5{,}698$$

EUAC = – *$4,782 or, simply an average annualized cost of $4,782.*

Figure 3-25 represents the equal uniform annual cost equivalent periodic cash flows corresponding to the cash flow scenario depicted in Figure 3-24.

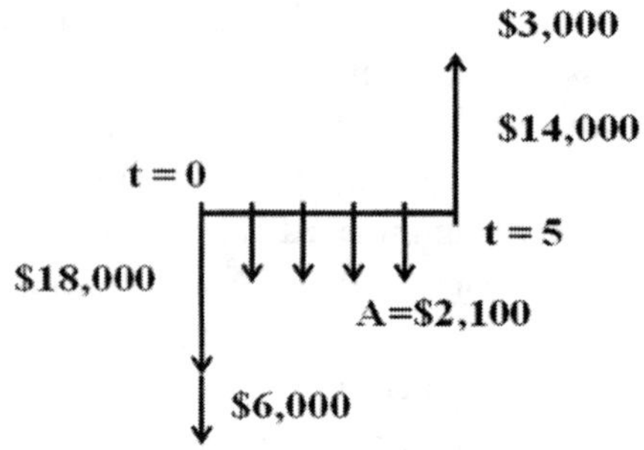

Figure 3-24

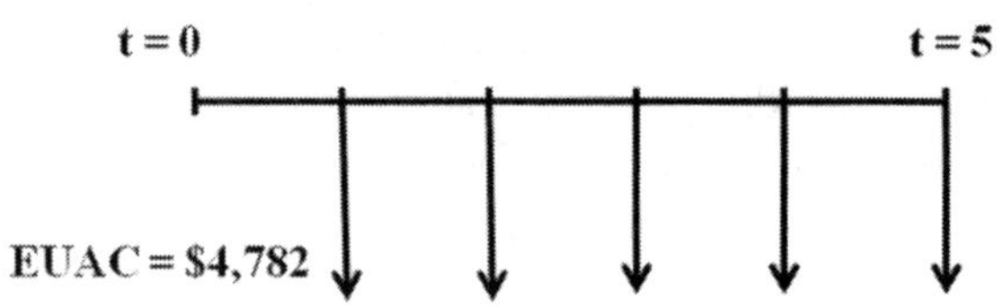

Figure 3-25

CASE STUDY 3-3: EUAC-BASED DECISION BETWEEN COMPETING ENERGY PROJECTS A AND B:

The particulars related to Projects A and B are listed in Table 3-1. The current discount rate in the market is 12%.

Objective: Evaluate the financials for the two alternative projects. Using the EUAC method, determine which of the two projects is more economical.

Solution:
EUAC Analysis,
Project A, Fuel Cell—Electrical Power Generator

$EUAC_A$ = – Annuity Equivalent of Initial Cost + Annuity Equivalent of the Future Salvage Value – Annual Preventive and Predictive Maintenance Cost – Annual Fuel Costs + Annual Energy Savings

Table 3-1

	Project A *Fuel Cell—Electrical* *Power Generator*	*Project B* *Diesel—Electrical* *Power Generator*
Initial, Turnkey, Installed Cost	$100,000	$60,000
Salvage Value in 20 Years	$10,000	$6,000
Annual Preventive and Predictive Maintenance Cost	$2,000	$5,000
Annual Fuel Costs	$1,200	$3,000
Annual Energy Savings	$30,000	$10,000

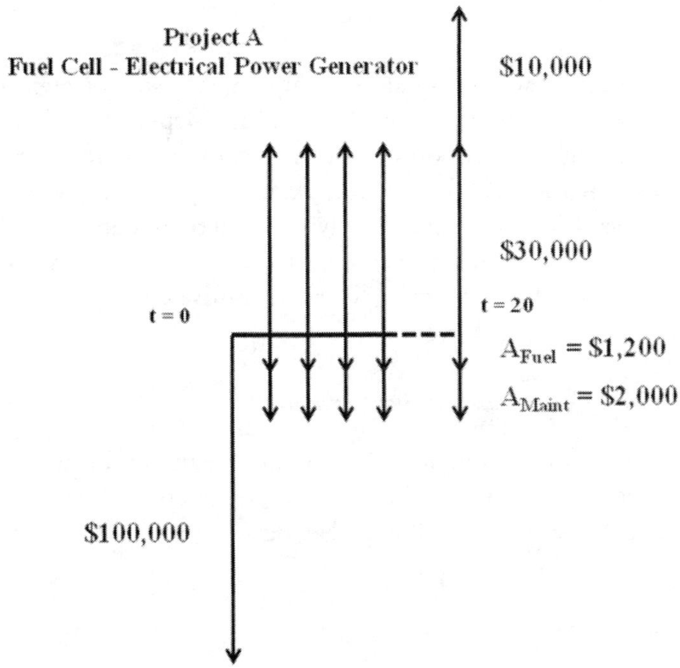

Figure 3-26

$$EUAC_A = -A_{Fuel} - A_{Maint} + A_{E\text{–}Savings} + F_{Salvage}\,(A/F,\ 12\%,\ 20) - P_{Initial\ Cost} \cdot (A/P,\ 12\%,\ 20)$$

$$EUAC_A = -\$1{,}200 - \$2{,}000 + \$30{,}000 + F_{Salvage}\,(0.0139) - P_{Initial\ Cost} \cdot (0.1339)$$

$$EUAC_A = -\$1{,}200 - \$2{,}000 + \$30{,}000 + (\$10{,}000) \cdot (0.0139) - (\$100{,}000) \cdot (0.1339)$$

$$EUAC_A = -\$1{,}200 - \$2{,}000 + \$30{,}000 + (\$139) - (\$13{,}390)$$

$$EUAC_A = +\ \$13{,}549$$

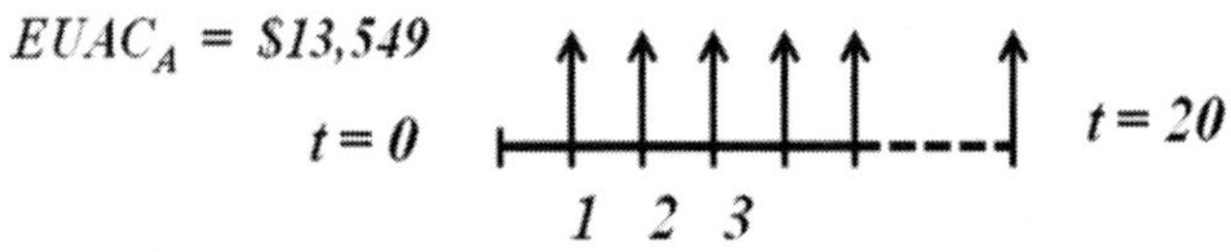

Figure 3-27

Figure 3-27 shows the equal uniform annual cost equivalent periodic cash flows corresponding to the cash flow scenario depicted in Figure 3-26.

Cash flow sign convention was adhered to in the solution for this case study. In other words, cost transactions were assigned a negative sign and positive sign was used for revenue or savings type transactions. The positive value for the EUAC, in this case study, signifies that Alternative A, largely due to the annual energy savings, will yield a positive cash flow for this firm.

EUAC Analysis,
Project B, Diesel—Electrical Power Generator

$EUAC_B$ = – Annuity Equivalent of Initial Cost + Annuity Equivalent of the Future Salvage Value – Annual Preventive and Predictive Maintenance Cost – Annual Fuel Costs + Annual Energy Savings

$$EUAC_B = -A_{Fuel} - A_{Maint} + A_{E\text{–}Savings} + F_{Salvage}\,(A/F,\ 12\%,\ 20) - P_{Initial\ Cost} \cdot (A/P,\ 12\%,\ 20)$$

$$EUAC_B = -\$3{,}000 - \$5{,}000 + \$10{,}000 + F_{Salvage} \cdot (0.0139) - P_{Initial\ Cost} \cdot (0.1339)$$

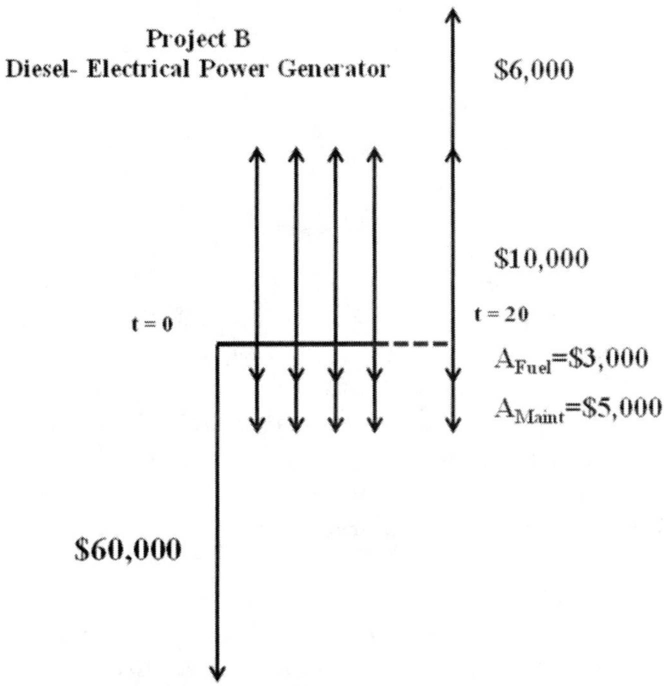

Figure 3-28

$$EUAC_B = -\$3{,}000 - \$5{,}000 + \$10{,}000 + (\$6{,}000) \cdot (0.0139) - (\$60{,}000) \cdot (0.1339)$$

$$EUAC_B = -\$5{,}951$$

Figure 3-29 shows the equal uniform annual cost equivalent periodic cash flows corresponding to Project B cash flow scenario depicted in Figure 3-28.

The negative value for the $EUAC_B$, shows that Alternative B, due to lower annual energy savings, will not yield any net savings.

Therefore, as obvious, Alternative A, fuel cell based electrical power generating system, catering to base load demand and yielding net annual cash flow of $13,549, on the basis of EUAC analysis, would be the preferred alternative.

Note that in the solution for Case Study 3-3, for the sake of maintaining congruence with the cash flow diagrams, all cost or negative cash flows

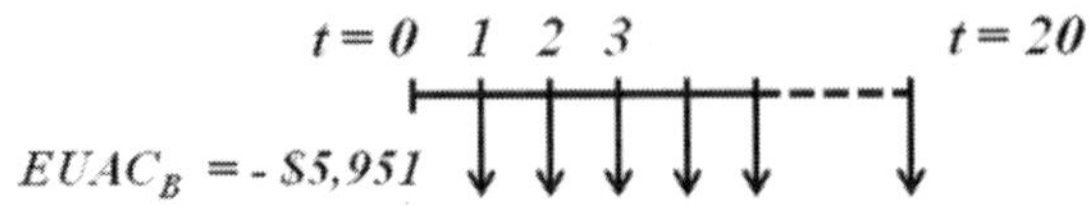

Figure 3-29

were treated as negative values and all positive cash flow type transactions were presented as positive values. Some texts and analysis methods treat the cost or expense type cash flows as positive cost values. The reasoning for the later school of thought's deviation from the traditional sign convention is that EUAC is a cost entity and is inherently negative. So, if EUAC computation includes traditionally positive cash flows, such as energy savings, in some realms it may be assigned a negative sign. If such an opposite convention had been employed in case study 3-3, the equation representing the solution strategy would have been as follows:

EUAC = + Annuity Equivalent of Initial Cost – Annuity Equivalent of the Future Salvage Value + Annual Preventive and Predictive Maintenance Cost + Annual Fuel Costs – Annual Energy Savings

COMPOUNDING VS. SIMPLE INTEREST

The concept of compounding represents the process of paying, receiving or accruing interest on not just the principal amount of funds but a combination of principal and interest.

For instance, if you invest $1,000 in a utility type investment fund today, at an annual interest or yield rate of 10%, you will have a compounded amount of $1,100 at the end of the first year. If you leave the funds in the investment account for the second year, the interest accrued in the second year will be computed on the basis of $1,100 investment, instead of the original amount of $1,000, amounting to a total of $1,210. This is the principle behind compounding and compound interest.

If, however, the interest is accrued only on the original, principal amount, it is said to be simple interest. In the previous example, the final amount at the end of 2 years would be $1,200, instead of $1,210.

Example 3-6

A principal amount of $1,000 is compounded monthly at a 6% nominal annual interest rate. How much will be accumulated in 5 years?

Solution:

If the given amount is compounded monthly at a nominal annual interest rate of 6%, the interest rate for the compounding period would be:

Eff. Int. Rate Per Compounding Period = ϕ =

$$\frac{\text{Nominal Rate Per Year "r"}}{\text{\# of Compounding Periods "k"}}$$

$\therefore \quad \phi$ = r/k
= Eff. Int. Rate Per Compounding Period
= 6%/12
= 0.5%

Where, n = Number of compounding periods
= 5 years x 12 months per year
= 60 months

$\therefore \quad F$ = P (F/P, 0.5%, 60)
= $1,000 x (1.3489)
= $1,348.90

SELF ASSESSMENT PROBLEMS AND QUESTIONS—CHAPTER 3

Problem 1

A small, privately held, power company is planning an expansion of capacity by 30 MW. This firm has decided to raise capital through acceptance of a few select partners. The firm's past performance shows an annual return of approximately 15%. If you were to invest $100,000 in this firm, what would investment be worth after 10 years? Hint: Use the rate of return as an interest rate.

A. $96,000
B. –$90,000
C. $404,600
D. $204,800

Problem 2

Assume that you invested $200,000 with the power company referenced in problem1, and 10 years later your investment has an estimated net worth of $1,000,000. What is the approximate rate of return on your investment? Hint: Use the time value of money equation method.

A. 20%
B. 10%
C. 12%
D. 17%

Problem 3

Assume that 5 years after you invest in the power company described in problem 1, the firm goes public and your shares in the publicly held firm is worth $500,000. According to your financial analyst, the new value of your investment translates into a 20% return. What was the value of your original investment in the firm? Hint: Use the rate of return as an interest rate.

A. $200,950
B. $100,000
C. $125,240
D. $340,000

Problem 4

If $1,000 is invested in a municipal bond investment portfolio today, $2,000 five years from now and $5,000 eight years from now, what would be the future value of this investment at the end of year 10 years? Assume an average discount rate 10%.

A. $8,543
B. $11,866

C. $10,000
D. $12,566

Problem 5

What would be the present value of investments in Problem 4 is:

A. $8,374
B. $5,375
C. $10,000
D. $4,574

Problem 6

If $1,000 is invested in a municipal bond investment portfolio on January 1, year 1, how much will be accumulated by January 1, year 12. The rate of return on the bond portfolio is 9%.

A. $1,580
B. $4,866
C. $2,580
D. $1,256

Problem 7

Two alternative (mutually exclusive) projects, X & Y, are being considered for expansion of an existing manufacturing facility. Both projects bear a life expectancy of 25 years. The salvage value, at the end of the expected life, is expected to be zero. The prevailing discount rate is 25%. The initial costs and average annualized costs of operation associated with each project are listed below:

	Project "X"	Project "Y"
Initial Cost	$22,000	$27,000
Annualized Cost of Operation	$3,100	$2,600

Using the NPV method, determine which project would cost the least?

A) Project X
B) Project Y

Problem 8

Using TVM (time value of money) analysis, determine which of the following options would yield higher future value, if the prevailing annual discount rate is 12%:

a) Option A: $100 invested each month for 24 months?
b) Option B: $1,200 invested each year, for 2 years?

Chapter 4

Financial Reporting Requirements

INTRODUCTION

The federal securities laws require companies with more than $10 million in assets and whose securities are held by more than 500 owners file annual and other periodic reports, regardless of whether the securities are ***publicly*** or ***privately*** traded. For example, domestic issuers (other than the small business issuers) must submit annual reports on Form 10-K, quarterly reports on Form 10-Q, and current reports on Form 8-K for a number of specified events. These reports must comply with certain disclosure requirements. These reporting requirements have been promulgated over time to protect the interest of the stockholders. These reporting requirements also serve as instruments for maintaining fairness, ethics and confidence in trading practices. For example, if 8-K reports were not required by the SEC, there would be little or no assurance that current or prospecting shareholders would be made aware of events or decisions that could impact their investments. At the same time, annual 10-K reports and the 10-Q reports, required by the SEC, ensure that investors are kept informed of fiscal health and well being of corporations, by corporate management, continually.

The reporting requirements are governed by the 2002 Sarbanes-Oxley Act of 2002. According to this Act, the CEO, chief executive officer, and CFO, chief financial officer, must certify the 10-K, 10-Q and 8-K filings by signing these reports. These reports must include current information about the control structure of the company.

The topic of SEC required financial reporting is important for energy engineers regardless of which side of the fence they happen to be. As an energy engineer on the client, customer, consumer or institution side of the fence, one needs to be aware of the financial standing of the supplier, contractor or vendor one is entering into contract with. These suppliers, contractors or vendors could be utility companies, equipment manufacturers, independent contractors, or engineering firms. These suppliers or contractors could be providing services, or implementing energy projects, through EPC or non-EPC programs. If these firms are large publically held corporations, you can obtain SEC reports of these firms to assess their fiscal standing.

If you happen to be an energy engineer or manager on the vendor or supplier side of the fence, you need to be aware of the financial health of your clients in the corporate world.

The annual report on Form 10-K provides a comprehensive overview of the company's business and financial condition and includes audited financial statements. Although similarly named, the annual report on Form 10-K is distinct from the "annual report to shareholders," which a company must send to its shareholders when it holds an annual meeting to elect directors.

Historically, Form 10-K had to be filed with the SEC within 90 days after the end of the company's fiscal year. However, in September 2002, the SEC approved a final rule that changed the deadlines to 75 days for Form 10-K and Form 10-Q for "accelerated filers"—meaning issuers that have a public float of at least $75 million, that have been subject to the Exchange Act's reporting requirements for at least 12 calendar months, that previously have filed at least one annual report, and that are not eligible to file their quarterly and annual reports on Forms 10-QSB and 10-KSB. These shortened deadlines will be phased in over time.

8-K REPORTS

Form 8-K is used to notify investors of a significant current event; an event that can impact the value of the firm's stock. Some examples of such events are listed below:

- Organizational Events:
  - Election of a new director.
  - A director leaves the firm. If a director departs because of a disagreement with the company on any matter relating to the registrant's operations, policies, or practices then an 8-K must be filed to disclose a brief description of the circumstances.

- — Changes in control of the company. Someone takes a large equity position (more than 15%); shareholder also needs to file with SEC as to intent. Someone takes a 51% position.
- — Changes in executive management. Officer departs, is fired or hired.
- — Change of accountant.
- — Amendments to company governance policies, such as, code of ethics and board committee governance policies.

- **Bankruptcies**
  —Bankruptcy of any type; i.e. Chapter 11 or Chapter 7.
  —Receiverships—a form of bankruptcy.

- **Business Agreements:**
  —Substantial and material, definitive, agreements that are not made in the ordinary course of business

- **Asset Acquisition or Sale:**
  —Substantial asset acquisitions, such as those associated with accretive growth or expansion of a firm.
  —Liquidation or sale of a substantial amount of assets, such as those associated with the divestures and "spin-offs."

- **Assumption of Debt or Significant Debt/Credit Related Events:**
  —Raising of capital through, material, direct financial obligations, i.e. bonds, debentures, etc. For example, funding of large energy projects through issuance of bonds.
  —Defaults on loans
  —Events that accelerate material obligations.
  —Change in credit or credit rating

- **Substantial Change in Business and Operating Conditions:**
  —Furloughs or layoffs, temporary or permanent.
  —Plant operations or services shut down
  —Substantial and material change in portfolio of services or distribution network
  —Material impairments
  —Substantial supply chain interruptions

- **Events Related to Stocks and Bonds, or Shareholder Equity or Firms' Financial Rating:**
  —Unregistered equity sales
  —Modifications to shareholder rights
  —Trading suspension

- **Legal Events or Actions**
  —SEC investigations and internal reviews
  —Financial non-reliance notices

- **Other Substantial or Material Events**

It is a good practice for investors to read any 8-K filings that are made by companies in which they are invested. These reports are both material and relevant to the company, and often contain information that will affect the share price.

Exhibit 4-1 illustrates the content and format of a typical 8-K report. This exhibit consists of a recent 8-K filing by Duke Energy.

Following observations are made upon examination of this Duke Energy 8-K report:

1. One of the events that triggered this 8-K filing, as recognized under the comment labeled "Item 5.02," is departure of certain directors and the election of certain directors.
2. Changes in the compensatory arrangements of certain executives.
3. Submission of certain matters to a vote of security holders. This included approval of Deloitte & Touche, LLP, as Duke Energy's independent public accountant for 2010.

Exhibit 4-1. Duke Energy 8-K Report filed on May 6, 2010.

UNITED STATES
SECURITIES AND EXCHANGE COMMISSION
Washington, D.C. 20549

FORM 8-K

CURRENT REPORT
Pursuant to Section 13 or 15(d) of the Securities Exchange Act of 1934

Date of Report (Date of earliest event reported): **May 6, 2010**

DUKE ENERGY CORPORATION
(Exact Name of Registrant as Specified in its Charter)

Delaware	**001-32853**	**20-2777218**
(State or Other Jurisdiction of Incorporation)	(Commission File Number)	(IRS Employer Identification No.)

526 South Church Street, Charlotte, North Carolina 28202
(Address of Principal Executive Offices, including Zip code)

(704) 594-6200
(Registrant's telephone number, including area code)

Check the appropriate box below if the Form 8-K filing is intended to simultaneously satisfy the filing obligation of the registrant under any of the following provisions:

☐ Written communications pursuant to Rule 425 under the Securities Act (17 CFR 230.425)

☐ Soliciting material pursuant to Rule 14a-12 under the Exchange Act (17 CFR 240.14a-12)

☐ Pre-commencement communications pursuant to Rule 14d-2(b) under the Exchange Act (17 CFR 240.14d-2(b))

☐ Pre-commencement communications pursuant to Rule 13e-4(c) under the Exchange Act (17 CFR 240. 13e-4(c))

Item 5.02. Departure of Directors or Certain Officers; Election of Directors; Appointment of Certain Officers; Compensatory Arrangements of Certain Officers.

On May 6, 2010, the shareholders of Duke Energy Corporation ("Duke Energy"), upon recommendation of our Board of Directors, approved the Duke Energy Corporation 2010 Long-Term Incentive Plan (the "2010 Plan"). A brief description of the 2010 Plan follows, but is subject to the full text of the plan attached as Appendix A to our proxy statement dated March 23, 2010.

The 2010 Plan authorizes the grant of equity-based compensation to our key employees and non-employee directors in the form of stock options, stock appreciation rights, restricted shares, performance shares, performance units, phantom stock, stock bonuses and dividend equivalents.

Duke Energy has reserved 75,000,000 shares of common stock for delivery under the 2010 Plan. The 2010 Plan contains a fungible share limit, which means that so-called "full value awards" such as restricted shares, performance awards, phantom stock, dividend equivalents paid in the form of Duke Energy common stock and stock bonuses are counted against the 75,000,000 share reserve as four shares for every one share that is delivered in connection with such award. As a result, up to 18,750,000 shares may be delivered as full value awards.

The 2010 Plan is administered by the Compensation Committee of the Board of Directors, which has authority to, among other things: construe and interpret the 2010 Plan, select participants and the types of awards to be granted, and establish the terms and conditions of awards. The Compensation Committee may grant performance awards that are intended to qualify for the "performance-based compensation" exemption under section 162(m) of the Internal Revenue Code, as well as performance awards that are not intended to so qualify. The performance criteria for a section 162(m) qualified award, which may relate to Duke Energy, any subsidiary, any business unit or any participant, and may be measured on an absolute or relative to peer group or other market measure basis, shall be limited to total shareholder return; stock price increase; return on equity; return on capital; earnings per share; EBIT (earnings before interest and taxes); EBITDA (earnings before interest, taxes, depreciation and amortization); ongoing earnings; cash flow (including operating cash flow, free cash flow, discounted cash flow return on investment, and cash flow in excess of costs of capital); EVA (economic value added); economic profit (net operating profit after tax, less a cost of capital charge); SVA (shareholder value added); revenues; net income; operating income; pre-tax profit margin; performance against business plan; customer service; corporate governance quotient or rating; market share; employee satisfaction; safety; reliability; employee engagement; supplier diversity; workforce diversity; operating margins; credit rating; dividend payments; expenses; operations and maintenance expenses; fuel cost per million BTU; costs per kilowatt hour; retained earnings; completion of acquisitions, divestitures and corporate restructurings; and individual goals based on objective business criteria underlying the goals listed above and which pertain to individual effort as to achievement of those goals or to one or more business criteria in the areas of litigation, human resources, information services, production, inventory, support services, site development, plant development, building development, facility development, government relations, product market share or management. In the case of a performance award that is not intended to qualify for exemption under section 162(m) of the Internal Revenue Code, the Compensation Committee shall designate performance criteria from among the foregoing or such other business criteria as it shall determine in its sole discretion.

The 2010 Plan will remain in effect until February 22, 2020, unless sooner terminated by the Board of Directors. Termination will not affect grants and awards then outstanding. The 2010 Plan replaces the Duke Energy Corporation 2006 Long-Term Incentive Plan, as amended (the "2006 Plan"). No further awards will be made under the 2006 Plan; however, awards granted under the 2006 Plan prior to shareholder approval of the 2010 Plan will remain outstanding in accordance with their terms.

Item 5.07. Submission of Matters to a Vote of Security Holders.

Duke Energy held its Annual Meeting of Shareholders on May 6, 2010. At the meeting, shareholders elected all 11 of the directors nominated by the Board of Directors. Each director received a greater number of votes cast "for" his or her election than votes cast "against" his or her election as reflected below. In addition, shareholders approved the Duke Energy Corporation 2010 Long-Term Incentive Plan and ratified the appointment of Deloitte & Touche LLP as Duke Energy's independent public accountant for 2010. Three shareholder proposals presented at the meeting were not approved. For more information on the proposals, see Duke Energy's proxy statement dated March 23, 2010. Set forth below are the final voting results for each of the proposals.

Election of Director Nominees

Director	Votes For	Votes Withheld	Abstentions	Broker Non-Votes
William Barnet, III	**753,303,222**	**16,538,017**	—	**317,987,250**
G. Alex Bernhardt, Sr.	748,179,557	21,661,682	—	317,987,250
Michael G. Browning	**747,386,033**	**22,455,206**	—	**317,987,250**
Daniel R. DiMicco	709,917,238	59,924,001	—	317,987,250
John H. Forsgren	**747,165,273**	**22,675,966**	—	**317,987,250**
Ann Maynard Gray	737,651,280	32,189,959	—	317,987,250
James H. Hance, Jr.	**735,929,951**	**33,911,288**	—	**317,987,250**
E. James Reinsch	754,600,409	15,240,830	—	317,987,250
James T. Rhodes	**754,785,602**	**15,055,637**	—	**317,987,250**
James E. Rogers	734,530,207	35,311,032	—	317,987,250
Philip R. Sharp	**754,168,447**	**15,672,792**	—	**317,987,250**

Proposal to approve the Duke Energy Corporation 2010 Long-Term Incentive Plan

Votes For	Votes Against	Abstentions	Broker Non-Votes
684,819,628	78,197,758	6,823,853	317,987,250

Proposal to ratify the appointment of Deloitte & Touche LLP as independent public accountant for 2010

Votes For	Votes Against	Abstentions	Broker Non-Votes
1,069,891,25513,260,222	4,536,534	0	

Shareholder proposal relating to preparation of a report on Duke Energy's global warming-related lobbying activities

Votes For	Votes Against	Abstentions	Broker Non-Votes
59,954,880	586,136,229	123,750,130	317,987,250

Shareholder Proposal regarding an amendment to our organizational documents to require majority voting for the election of directors

Votes For	Votes Against	Abstentions	Broker Non-Votes
309,434,594	452,725,011	7,681,634	317,987,250

Shareholder Proposal regarding the adoption of a policy requiring senior executives to hold a portion of their equity grants until after retirement.

Votes For	Votes Against	Abstentions	Broker Non-Votes
186,148,678	571,575,510	12,117,051	317,987,250

SIGNATURE

Pursuant to the requirements of the Securities and Exchange Act of 1934, the registrant has duly caused this report to be signed on its behalf by the undersigned hereunto duly authorized.

DUKE ENERGY CORPORATION

Date: May 12, 2010

By: /s/Marc E. Manly

Name: Marc E. Manly

Title: Group Executive, Chief Legal Officer and Corporate Secretary

10-K REPORTS

Form 10-K is used to file an annual report required by the U.S. Securities and Exchange Commission (SEC). This report provides a comprehensive summary of a public company's performance. This report is different from the "annual report to shareholders," which a company must send to its shareholders when it holds an annual meeting to elect directors. However, some firms do combine the annual report and the 10-K into one document. The 10-K includes information such as company history, organizational structure, executive compensation, equity, subsidiaries, and audited financial statements. If a shareholder requests a company's Form 10-K, the company must provide a copy. Until March 16, 2009, smaller companies could use Form 10-KSB.

Exhibit 4-2 shows Duke Energy's 10-K report for year 2010. This exhibit illustrates the content and format of a typical 10-K report. Following observations can be made upon examination of this Duke Energy 10-K report:

1. This report consists of a summary of material information on Dukes Energy's financial performance in 2009. This information includes the following:

   a. For the year-ended December 31, 2009, Duke Energy Corporation (Duke Energy) reported net income, attributable to Duke Energy, of $1,075 million.

   b. Basic and diluted earnings per share (EPS) of $0.83.

   c. Income from continuing operations was $1,073 million for 2009 as compared to $1,275 million for 2008.

   d. Total reportable segment EBIT, Earnings Before Income Tax decreased to $2,713 million in 2009 from $3,073 million in 2008.

2. Other information included in this 10-K report, that is material and vital for investors, highlights the competitive pressures encountered by Duke in 2009. The report outlines measures adopted by Duke to overcome competitive challenges and to maximize utilization of opportunities.

As recommend earlier for 8-K reports, it is a good practice for current and prospective investors to read 10-K filings that are made by companies. These reports are both material and relevant to the company, and often contain information that can answer investor's questions and concerns.

Exhibit 4-2. Duke Energy 10-K Report, 2010

UNITED STATES SECURITIES AND EXCHANGE COMMISSION
WASHINGTON, D.C. 20549

FORM 10-K

FOR ANNUAL AND TRANSITION REPORTS PURSUANT TO SECTION 13 OR 15(d) OF THE SECURITIES EXCHANGE ACT OF 1934

(Mark One)

☒ ANNUAL REPORT PURSUANT TO SECTION 13 OR 15(d) OF THE SECURITIES EXCHANGE ACT OF 1934
For the fiscal year ended December 31, 2009 or

☐ TRANSITION REPORT PURSUANT TO SECTION 13 OR 15(d) OF THE SECURITIES EXCHANGE ACT OF 1934
For the transition period from _______ to _______

Commission file number 1-32853

DUKE ENERGY CORPORATION

(Exact name of registrant as specified in its charter)

Delaware	20-2777218
(State or other jurisdiction of incorporation or organization)	(I.R.S. Employer Identification No.)
526 South Church Street, Charlotte, North Carolina	28202-1803
(Address of principal executive offices)	(Zip Code)

704-594-6200
(Registrant's telephone number, including area code)

SECURITIES REGISTERED PURSUANT TO SECTION 12(B) OF THE ACT:

Title of each class	Name of each exchange on which registered
Common Stock, $0.001 par value	New York Stock Exchange, Inc.

Indicate by check mark if the registrant is a well-known seasoned issuer, as defined in Rule 405 of the Securities Act. Yes ☒ No ☐

Indicate by check mark if the registrant is not required to file reports pursuant to Section 13 or Section 15(d) of the Exchange Act. Yes ☐ No ☒

Indicate by check mark whether the registrant (1) has filed all reports required to be filed by Section 13 or 15(d) of the Securities Exchange Act of 1934 during the preceding 12 months (or for such shorter period that the registrant was required to file such reports) and (2) has been subject to such filing requirements for the past 90 days. Yes ☒ No ☐

Indicate by check mark whether the registrant has submitted electronically and posted on its corporate Web site, if any, every Interactive Data File required to be submitted and posted pursuant to Rule 405 of Regulation S-T (§232.405 of this chapter) during the preceding 12 months (or for such shorter period that the registrant was required to submit and post such files). Yes ☒ No ☐

Indicate by check mark if disclosure of delinquent filers pursuant to Item 405 of Regulation S-K is not contained herein, and will not be contained, to the best of registrant's knowledge, in definitive proxy or information statements incorporated by reference in Part III of this Form 10-K or any amendment to this Form 10-K. ☐

Indicate by check mark whether the registrant is a large accelerated filer, an accelerated filer, a non-accelerated filer, or a smaller reporting company. See the definitions of "large accelerated filer," "accelerated filer" and "smaller reporting company" in Rule 12b-2 of the Exchange Act. (Check one):

Large accelerated filer ☒	Accelerated filer ☐
Non-accelerated filer ☐	Smaller reporting company ☐

(Do not check if a smaller reporting company)

Indicate by check mark whether the registrant is a shell company (as defined in Rule 12b-2 of the Securities Exchange Act of 1934). Yes ☐ No ☒

Estimated aggregate market value of the common equity held by nonaffiliates of the registrant at June 30, 2009

$18,836,000,000

Number of shares of Common Stock, $0.001 par value, outstanding at February 22, 2010.

1,309,314,484

Form 10-K for DUKE ENERGY CORP

26-Feb-2010
Annual Report—Summary

Item 7. Management's Discussion and Analysis of Financial Condition and Results of Operations.

INTRODUCTION

Management's Discussion and Analysis should be read in conjunction with the Consolidated Financial Statements and Notes for the years ended December 31, 2009, 2008 and 2007.

EXECUTIVE OVERVIEW

2009 Financial Results. For the year-ended December 31, 2009, Duke Energy Corporation (Duke Energy) reported net income attributable to Duke Energy of $1,075 million and basic and diluted earnings per share (EPS) of $0.83, as compared to net income attributable to Duke Energy of $1,362 million and basic and diluted EPS of $1.08 and $1.07, respectively, for the year-ended December 31, 2008. Income from continuing operations was $1,073 million for 2009 as compared to $1,275 million for 2008. Total reportable segment EBIT (defined below in "Segment Results" section of Management's Discussion and Analysis of Financial Condition and Results of Operations) decreased to $2,713 million in 2009 from $3,073 million in 2008.

See "Results of Operations" below for a detailed discussion of the consolidated results of operations, as well as a detailed discussion of EBIT results for each of Duke Energy's reportable business segments, as well as Other.

2009 Areas of Focus and Accomplishments. In 2009, management was focused on managing through the economic recession, investing in modernization of Duke Energy's regulated infrastructure and dealing with increased competition in Ohio.

Managing Through the Economic Recession and Changing Competitive Landscapes. In U.S. Franchised Electric and Gas, Duke Energy's largest business segment, weather-normalized electric volumes were down approximately 4% when compared to 2008. This was driven primarily by a decrease in industrial sales volumes, which were down approximately 14% compared to 2008. Although industrial sales volumes were down year over year, industrial volumes began to show signs of stabilization late in 2009. On a weather-normalized basis, residential sales volumes were slightly positive, while commercial sales volumes were slightly negative. Looking forward to 2010, management expects the load forecast to be relatively flat compared to 2009.

In 2009, Commercial Power's operations were impacted by the competitive markets in Ohio, which were triggered by low commodity prices that put downward pressure on power prices. The available capacity and lower prices provided opportunities for native load customers in Ohio to switch generation suppliers. Competitive power suppliers began supplying power to current Commercial Power native load customers in Ohio and Commercial Power experienced an increase in customer switching beginning in the second quarter of 2009. As of December 31, 2009, customer switching levels approximated 40% of Commercial Power's native load. However, through Duke Energy Retail Sales (DERS), Commercial Power acquired approximately 60% of the switched load by offering customers a discount to the Electric Security Plan (ESP) price. When factoring in the DERS activity, Commercial Power experienced net customer switching of about 15%, although those native load customers acquired by DERS were at lower margins than customers served under the ESP. Additionally, DERS has been able to acquire new customers outside Commercial Power's native load territory. As a result of lower forecasted energy prices, lower demand for electricity due to the economy and competitive pressures in Ohio, and other valuation factors, a non-cash goodwill impairment charge of approximately $371 million was recorded by Commercial Power in the third quarter of 2009.

In light of the above economic factors that impacted Duke Energy's business in 2009, management was focused on offsetting those economic pressures by successfully managing costs and achieving excellent operational performance. Duke Energy achieved significant operations and maintenance cost mitigation goals across its business segments and also reduced planned capital expenditures by approximately $200 million, which highlights Duke Energy's ability to take advantage of the flexibility within its capital spending plan. Additionally, Duke Energy's generation fleet operated at some of the highest levels in Duke Energy's history. These combined efforts allowed Duke Energy to largely mitigate the negative impact of the economy on its results of operations in 2009.

Key Regulatory Accomplishments. During 2009, Duke Energy completed the following regulatory initiatives:

• Obtained favorable rate case outcomes in North Carolina, South Carolina, Ohio and Kentucky which will increase revenues by nearly $460 million upon full implementation.

• Updated/enabled construction work-in-progress (CWIP) recovery for Duke

Energy Carolinas' Cliffside Unit 6 and the Integrated Gasification Combined Cycle (IGCC) plant at Duke Energy Indiana's Edwardsport Generating Station.

• Received approval for cost recovery mechanisms for save-a-watt programs in North Carolina, South Carolina and Ohio. Approval in Indiana is anticipated in February 2010.

• Began deployment of SmartGrid in Ohio, along with the initiation of a rate rider cost recovery mechanism, which is awaiting approval and a ruling is expected in the first quarter of 2010. Additionally, Duke Energy was awarded a stimulus grant for approximately $200 million to be used for reimbursement of costs related to SmartGrid.

• Received approvals of wind, solar and other renewable energy projects, which will enable innovative renewable energy initiatives and help Duke Energy meet specific renewable energy standards over time.

Overall, the regulatory and legislative accomplishments during 2009 have positioned Duke Energy well for 2010 and beyond.

Capital Expenditures and Fleet and Grid Modernization. Duke Energy's strategy for meeting customer demand, while building a sustainable business that allows its customers and its shareholders to prosper in a carbon-constrained environment, includes significant commitments to renewable energy, customer energy efficiency, advanced nuclear power, advanced clean-coal and high-efficiency natural gas electric generating plants, and retirement of older less efficient coal-fired power plants. Due to the likelihood of upcoming environmental regulations, including carbon legislation, air pollutant regulation by the U.S. Environmental Protection Agency (EPA) and coal regulation, Duke Energy has been focused on modernizing its fleet in preparation for a low carbon future. During 2009, Duke Energy has continued the construction of Cliffside Unit 6 in North Carolina and the Edwardsport IGCC plant in Indiana and these construction projects are approximately 55% complete and 50% complete, respectively, at December 31, 2009. Both are scheduled to be placed in service during 2012. Once in service, Duke Energy will begin retiring older, less efficient coal and gas-fired units. Additionally, Duke Energy Carolinas has begun construction on a 620 megawatt (MW) combined cycle natural gas-fired generating facility at each of its existing Buck and Dan River Steam Stations. These facilities are scheduled to be placed in service in 2011 and 2012, respectively. In conjunction with these and other capital projects, management is continuing its focus on reducing regulatory lag, which refers to the period of time between making an investment and earning a return and recovering that investment. In 2007, the Indiana Utility Regulatory Commission (IURC) approved the timely recovery of initial construction cost estimates associated with the Edwardsport IGCC plant. The 2009 rate case settlements in North Carolina and South Carolina included stipulations allowing for the recovery in base rates of financing costs related to Cliffside Unit 6, although the recovery is delayed in North Carolina for a one year period.

Table of Contents
PART II

Duke Energy Carolinas is also continuing to seek all necessary regulatory approvals for the proposed William States Lee III Nuclear Station, including the December 2007 filings of a Combined Construction and Operating License (COL) application with the Nuclear Regulatory Commission (NRC) and requests to incur up to $230 million in development costs through 2009, which were approved in 2008. Although these actions are necessary steps as management continues to pursue the option of building a new nuclear plant, submitting these applications does not commit Duke Energy Carolinas to build a nuclear unit.

In 2009, Duke Energy made significant strides in adding to its existing renewable energy portfolio. One way Duke Energy is reducing its environmental footprint while meeting demand for reliable, clean energy is by investing in zero carbon wind power. During 2009, Commercial Power, through Duke Energy Generation Services (DEGS), brought approximately 364 MW of wind generation online through a combination of completed construction and acquisition. At December 31, 2009, DEGS had approximately 735 MW of wind generation in commercial operation. The wind assets in service have long-term power purchase agreements to sell the output to an end customer. Additionally, DEGS became an owner in a biomass development joint venture and, in early 2010, announced it would acquire a 16 MW solar development project in San Antonio, Texas.

Management is also making progress on increasing the role energy efficiency will have in meeting customers' growing energy needs. Energy efficiency is considered a "fifth fuel" in the portfolio available to meet customers' growing needs for electricity, along with coal, nuclear, natural gas and renewable energy. During 2009, Duke Energy's save-a-watt models were approved in North Carolina, South Carolina and Ohio and Duke Energy is awaiting a decision on the proposed save-a-watt model in Indiana, which is expected in the first quarter of 2010. The save-a-watt proposal in Kentucky was withdrawn and will be addressed in Duke Energy Kentucky's next general rate case.

Duke Energy Objectives—2010 and beyond. Duke Energy will continue to focus on operational excellence, shaping federal and state legislative and regulatory policy, continued modernization of infrastructure and investing in renewable energy, including energy efficiency. The majority of future earnings are anticipated to be contributed from U.S. Franchised Electric and Gas, which consists of Duke Energy's regulated businesses that currently own a capacity of approximately 27,000 MW of generation. The regulated generation portfolio consists of a mix of coal, nuclear, natural gas and hydroelectric generation, with the substantial majority of all of the sales of electricity coming from coal and nuclear generation facilities. The favorable rate case outcomes reached in the various jurisdictions in 2009, as discussed above, will increase U.S. Franchised Electric and Gas' revenues by approximately $460 million upon full implementation.

As a result of the downturn in the economy, Duke Energy experienced reductions in sales volumes in 2009, most notably within the industrial customer class. Management anticipates that recessionary pressures will continue in 2010, resulting in essentially flat kilowatt-hour sales in both the Carolinas and the Midwest service territories. In order to address these pressures, management is focused on containing costs in 2010 and currently expects non-recoverable (i.e., not directly recovered via a rider or other mechanism) operations and maintenance expense to be flat compared to 2009, due largely to sustainable reductions achieved during 2009, as well as certain 2010 initiatives such as a voluntary severance program and office consolidation. In addition, management will continue efforts to achieve constructive regulatory outcomes to reduce regulatory lag, including continually reviewing the need for general rate case filings in certain jurisdictions in 2010 and beyond.

Additionally, due to the competitive markets in Ohio, customer switching will continue to impact the results of the Commercial Power business, as management currently estimates that an incremental 5% of current customer load will switch to alternative suppliers in 2010. Management is focused on mitigating lost volume and margin erosion in 2010 through DERS efforts to acquire native load customers, as well as acquiring customers outside of Commercial Power's Ohio native load territory that are currently supplied by other electric generators.

During the three-year period from 2010 through 2012, Duke Energy anticipates total capital expenditures of approximately $14 billion to $15 billion. Of this amount, approximately $5.7 billion is expected to be spent on committed projects, including base load power plants to meet long-term growth in customer demand and to modernize the generation fleet, ongoing environmental projects, and nuclear fuel. Approximately $6.8 billion of capital expenditures are expected to be used primarily for overall system maintenance, customer connections, and corporate expenditures. Although these expenditures are ultimately necessary to ensure overall system maintenance and reliability, the timing of the expenditures may be influenced by broad economic conditions and customer growth. The remaining estimated capital expenditures of approximately $1.2 billion to $2.7 billion are of a discretionary nature and relate to growth opportunities in which Duke Energy may invest, provided there are opportunities to meet return expectations along with assurance of constructive regulatory treatment in the regulated businesses. Discretionary capital primarily includes Commercial Power renewable and transmission projects, projects at International Energy and renewable projects at U.S. Franchised Electric and Gas. Capital expenditures are currently estimated to be approximately $5.2 billion in 2010. These expenditures are principally related to expansion plans, maintenance costs, environmental spending related to Clean Air Act (CAA) requirements and nuclear fuel. Duke Energy is committed to adding base load capacity at a reasonable price while modernizing the current generation facilities by replacing older, less efficient plants with cleaner, more efficient plants. Significant expansion projects include the Edwardsport IGCC plant, an 825 MW coal unit at Duke Energy Carolinas' existing Cliffside facility and new gas-fired generation units at Duke Energy Carolinas' existing Dan River and Buck Steam Stations, as well as other additions due to system growth. Additionally, Duke Energy is evaluating the potential construction of the William States Lee III nuclear power plant in Cherokee County, South Carolina.

Duke Energy anticipates capital expenditures at Commercial Power will primarily relate to growth opportunities, such as renewable energy generation projects and environmental control equipment, as well as maintenance on existing plants. Capital expenditures at International Energy, which will be funded with cash held or raised by International Energy, will primarily be for strategic growth opportunities, as well as maintenance on existing plants.

With the exception of equity issuances to fund the dividend reinvestment plan and other internal plans, Duke Energy does not currently anticipate the issuance of any other common equity in the foreseeable future. Duke Energy expects to have access to liquidity in the capital markets at reasonable rates and terms in 2010. Additionally, Duke Energy has access to unsecured revolving credit facilities, which are not restricted upon general market conditions, with aggregate bank commitments of approximately $3.14 billion. At December 31, 2009, Duke Energy has available borrowing capacity of approximately $1.9 billion under this facility. For further information related to management's assessment of liquidity and capital resources, including known trends and uncertainties, see "Liquidity and Capital Resources" below.

As the majority of Duke Energy's anticipated future capital expenditures are related to its regulated operations, a risk to Duke Energy is the ability to recover costs related to such expansion in a timely manner. Energy legislation passed in North Carolina and South Carolina in 2007 provides, among other things, mechanisms for Duke Energy to recover financing costs for new nuclear or coal base load generation during the construction phase. In Indiana, Duke Energy has received approval to recover its development costs for the new IGCC plant at the Edwardsport Generating Station. Duke Energy has received approval for nearly $260 million of future federal tax credits related to costs to be incurred for the modernization of Cliffside Unit 6, as well as the IGCC plant in Indiana. In addition, Duke Energy has received general assurances from the North Carolina Utilities Commission (NCUC) that the North Carolina allocable portion of development costs associated with the William States Lee III nuclear station will be recoverable through a future rate case proceeding as long as the costs are deemed prudent and reasonable. Duke Energy does not anticipate beginning construction of the proposed nuclear power plant without adequate assurance of cost recovery from the state legislators or regulators.

Table of Contents
PART II

In summary, Duke Energy is coordinating its future capital expenditure requirements with regulatory initiatives in order to ensure adequate and timely cost recovery while continuing to provide low cost energy to its customers.

Economic Factors for Duke Energy's Business. Duke Energy's business model provides diversification between stable regulated businesses like U.S. Franchised Electric and Gas and certain portions of Commercial Power's operations, and the traditionally higher-growth businesses like the unregulated portion of Commercial Power's operations and International Energy. As was the case throughout much of 2009, all of Duke Energy's businesses can be negatively affected by sustained downturns or sluggishness in the economy, including low market prices of commodities, all of which are beyond Duke Energy's control, and could impair Duke Energy's ability to meet its goals for 2010 and beyond.

As Duke Energy experienced in 2009, declines in demand for electricity as a result of economic downturns reduce overall electricity sales and have the potential to lessen Duke Energy's cash flows, especially as industrial customers reduce production and, thus, consumption of electricity. A weakening economy could also impact Duke Energy's customer's ability to pay, causing increased delinquencies, slowing collections and lead to higher than normal levels of accounts receivables, bad debts and financing requirements. A portion of U.S. Franchised Electric and Gas' business risk is mitigated by its regulated allowable rates of return and recovery of fuel costs under fuel adjustment clauses. The ESP in Ohio also helps mitigate a portion of the risk associated with certain portions of Commercial Power's generation operations by providing mechanisms for recovery of certain costs associated with, among other things, fuel and purchased power for native-load customers.

If negative market conditions should persist over time and estimated cash flows over the lives of Duke Energy's individual assets, including goodwill, do not exceed the carrying value of those individual assets, asset impairments may occur in the future under existing accounting rules and diminish results of operations. A change in management's intent about the use of individual assets (held for use versus held for sale) could also result in impairments or losses.

Duke Energy's 2010 goals can also be substantially at risk due to the regulation of its businesses. Duke Energy's businesses in the United States (U.S.) are subject to regulation on the federal and state level. Regulations, applicable to the electric power industry, have a significant impact on the nature of the businesses and the manner in which they operate. New legislation and changes to regulations are ongoing, including anticipated carbon legislation, and Duke Energy cannot predict the future course of changes in the regulatory or political environment or the ultimate effect that any such future changes will have on its business.

Duke Energy's earnings are impacted by fluctuations in commodity prices. Exposure to commodity prices generates higher earnings volatility in the unregulated businesses as there are timing differences as to when such costs are recovered in rates. To mitigate these risks, Duke Energy enters into derivative instruments to effectively hedge some, but not all, known exposures.

Additionally, Duke Energy's investments and projects located outside of the United States expose Duke Energy to risks related to laws of other countries, taxes, economic conditions, fluctuations in currency rates, political conditions and policies of foreign governments. Changes in these factors are difficult to predict and may impact Duke Energy's future results.

Duke Energy also relies on access to both short-term money markets and longer-term capital markets as a source of liquidity for capital requirements not met by cash flow from operations. An inability to access capital at competitive rates or at all could adversely affect Duke Energy's ability to implement its strategy. Market disruptions or a downgrade of Duke Energy's credit rating may increase its cost of borrowing or adversely affect its ability to access one or more sources of liquidity. Additionally, there are no assurances that commitments made by lenders under Duke Energy's credit facilities will be available if needed as a source of funding due to ongoing uncertainties in the financial services industry.

For further information related to management's assessment of Duke Energy's risk factors, see Item 1A. "Risk Factors."

RESULTS OF OPERATIONS

Consolidated Operating Revenues

Year Ended December 31, 2009 as Compared to December 31, 2008. Consolidated operating revenues for 2009 decreased approximately $476 million compared to 2008. This change was primarily driven by the following:

- An approximate $726 million decrease at U.S. Franchised Electric and Gas. See Operating Revenue discussion within "Segment Results" for U.S. Franchised Electric and Gas below for further information; and

- An approximate $27 million decrease at International Energy. See Operating Revenue discussion within "Segment Results" for International Energy below for further information.

Partially offsetting these decreases was:

- An approximate $288 million increase at Commercial Power. See Operating Revenue discussion within "Segment Results" for Commercial Power below for further information.

Year Ended December 31, 2008 as Compared to December 31, 2007. Consolidated operating revenues for 2008 increased approximately $487 million compared to 2007. This change was primarily driven by the following:

- An approximate $419 million increase at U.S. Franchised Electric and Gas. See Operating Revenue discussion within "Segment Results" for U.S. Franchised Electric and Gas below for further information; and

- An approximate $125 million increase at International Energy. See Operating Revenue discussion within "Segment Results" for International Energy below for further information.

Partially offsetting these increases was:

- An approximate $55 million decrease at Commercial Power. See Operating Revenue discussion within "Segment Results" for Commercial Power below for further information.

Consolidated Operating Expenses

Year Ended December 31, 2009 as Compared to December 31, 2008. Consolidated operating expenses for 2009 decreased approximately $247 million compared to

2008. This change was driven primarily by the following:

- An approximate $626 million decrease at U.S. Franchised Electric and Gas. See Operating Expense discussion within "Segment Results" for U.S. Franchised Electric and Gas below for further information;

- An approximate $65 million decrease at International Energy. See Operating Expense discussion within "Segment Results" for International Energy below for further information; and

- An approximate $40 million decrease at Other. See Operating Expense discussion within "Segment Results" for Other below for further information.

Table of Contents
PART II

Partially offsetting these decreases was:

- An approximate $489 million increase at Commercial Power, which includes approximately $413 million of impairment charges in 2009 primarily related to a goodwill impairment charge associated with the non-regulated generation operations in the Midwest. See Operating Expense discussion within "Segment Results" for Commercial Power below for further information.

Year Ended December 31, 2008 as Compared to December 31, 2007. Consolidated operating expenses for 2008 increased approximately $543 million compared to 2007. This change was driven primarily by the following:

- An approximate $401 million increase at U.S. Franchised Electric and Gas. See Operating Expense discussion within "Segment Results" for U.S. Franchised Electric and Gas below for further information;

- An approximate $123 million increase at International Energy. See Operating Expense discussion within "Segment Results" for International Energy below for further information; and

- An approximate $27 million increase at Commercial Power. See Operating Expense discussion within "Segment Results" for Commercial Power below for further information.

Consolidated Gains (Losses) on Sales of Other Assets and Other, net

Consolidated gains (losses) on sales of other assets and other, net was a gain of approximately $36 million and $69 million in 2009 and 2008, respectively, and a loss of approximately $5 million for 2007. The gains and losses for all years relate primarily to sales of emission allowances by U.S. Franchised Electric and Gas and Commercial Power.

Consolidated Operating Income

Year Ended December 31, 2009 as Compared to December 31, 2008. For 2009, consolidated operating income decreased approximately $262 million compared to 2008. Drivers to operating income are discussed above.

Year Ended December 31, 2008 as Compared to December 31, 2007. For 2008, consolidated operating income increased approximately $18 million compared to 2007. Drivers to operating income are discussed above.

Other drivers to operating income are discussed above. For more detailed discussions, see the segment discussions that follow.

Consolidated Other Income and Expenses

Year Ended December 31, 2009 as Compared to December 31, 2008. For 2009, consolidated other income and expenses increased approximately $212 million compared to 2008. This increase was primarily driven by an increase in equity earnings of approximately $172 million due mostly to impairment charges recorded by Crescent JV (Crescent) in 2008, of which Duke Energy's proportionate share was approximately $238 million, partially offset by decreased equity earnings from International Energy of approximately $55 million primarily related to lower contributions from its investment in National Methanol Company (NMC) and losses from its investment in Attiki Gas Supply S.A. (Attiki). Also, the mark-to-market and investment income on investments that support benefit obligations and within the captive insurance portfolio increased approximately $45 million as a result of gains in 2009 compared to losses in 2008. Additionally, foreign exchange impacts, primarily related to the remeasurement of certain U.S. dollar denominated cash and debt balances at International Energy, resulted in gains in 2009 compared to losses in 2008 due to favorable foreign exchange rates, resulting in an increase of approximately $43 million in 2009 compared to 2008. Partially offsetting these increases was decreased interest income of approximately $53 million due primarily to lower average cash and short-term investment balances, an approximate $26 million charge in 2009 related to certain performance guarantees Duke Energy had issued.

10-Q REPORT

Form 10-Q is used to file a quarterly report required by the U.S. Securities and Exchange Commission (SEC). This quarterly report requirement is pursuant to Section 13 or 15(d) of the Securities Exchange Act of 1934. This report contains information similar to the annual form 10-K, however the information is less detailed and the financial statements are generally unaudited.

Exhibit 4-3 shows Duke Energy's 10-Q report for First Quarter, 2010.

Exhibit 4-3. Duke Energy 10-Q Report filed on March 31, 2010.

UNITED STATES
SECURITIES AND EXCHANGE COMMISSION
WASHINGTON, D.C. 20549

FORM 10-Q

(Mark One)

☒ **QUARTERLY REPORT PURSUANT TO SECTION 13 OR 15(d) OF THE SECURITIES EXCHANGE ACT OF 1934**
For the quarterly period ended March 31, 2010

Or

☐ **TRANSITION REPORT PURSUANT TO SECTION 13 OR 15(d) OF THE SECURITIES EXCHANGE ACT OF 1934**
For the transition period from _____ to _____

Commission file number	Exact name of registrants as specified in their charters, addresses of principal executive offices, charters, addresses of principal executive offices, telephone numbers and states of incorporation	IRS Employer Identification No.

1-32853 DUKE ENERGY CORPORATION 20-2777218
526 South Church Street
Charlotte, NC 28202-1803
704-594-6200
State of Incorporation: Delaware

1-4928 DUKE ENERGY CAROLINAS, LLC 56-0205520
526 South Church Street
Charlotte, NC 28202-1803
704-594-6200
State of Incorporation: North 1-1232 D U K E
ENERGY OHIO, INC. 31-0240030
139 East Fourth Street
Cincinnati, OH 45202
704-594-6200
State of Incorporation: Ohio

1-3543 DUKE ENERGY INDIANA, INC. 35-0594457
1000 East Main Street
Plainfield, IN 46168
704-594-6200
State of Incorporation: Indiana

Indicate by check mark whether the registrant (1) has filed all reports required to be filed by Section 13 or 15(d) of the Securities Exchange Act of 1934 during the preceding 12 months (or for such shorter period that the registrant was required to file such reports), and (2) has been subject to such filing requirements for the past 90 days.

Duke Energy Corporation (Duke Energy)	Yes ☒	No ☐
Duke Energy Carolinas, LLC (Duke Energy Carolinas)	Yes ☒	No ☐
Duke Energy Ohio, Inc. (Duke Energy Ohio)	Yes ☒	No ☐
Duke Energy Indiana, Inc. (Duke Energy Indiana)	Yes ☒	No ☐

Indicate by check mark whether the registrant has submitted electronically and posted on its corporate Web site, if any, every Interactive Data File required to be submitted and posted pursuant to Rule 405 of Regulation S-T (§232.405 of this chapter) during the preceding 12 months (or for such shorter period that the registrant was required to submit and post such files).

Duke Energy	Yes ☒	No ☐
Duke Energy Carolinas	Yes ☒	No ☐

Duke Energy Ohio	Yes	☒	No	☐
Duke Energy Indiana	Yes	☒	No	☐

Indicate by check mark whether the registrant is a large accelerated filer, an accelerated filer, a non-accelerated filer, or a smaller reporting company. See the definitions of "large accelerated filer," "accelerated filer" and "smaller reporting company" in Rule 12b-2 of the Exchange Act. (Check one):

Duke Energy	Large accelerated filer	☒	Accelerated filer	☐
	Non-accelerated filer	☐	Smaller reporting company	☐
Duke Energy Carolinas	Large accelerated filer	☐	Accelerated filer	☐
Non-accelerated filer		☒	Smaller reporting company	☐
Duke Energy Ohio	Large accelerated filer	☒	Accelerated filer	☐
	Non-accelerated filer	☐	Smaller reporting company	☐
Duke Energy Indiana	Large accelerated filer	☐	Accelerated filer	☐
	Non-accelerated filer	☒	Smaller reporting company	☐

Indicate by check mark whether the registrant is a shell company (as defined in Rule 12b-2 of the Exchange Act).

Duke Energy	Yes	_	No	☒
Duke Energy Carolinas	**Yes**	☐	No	☒
Duke Energy Ohio	**Yes**	☐	No	☒

Table of Contents

Duke Energy Indiana	Yes	☐	No	☒

Indicate the number of shares outstanding of each of the Issuer's classes of common stock, as of the latest practicable date.

Outstanding as of May 3, 2010

Registrant	**Description**	**Shares**
Duke Energy	Common Stock, par value $0.001	1,313,134,663

Duke Energy Carolinas	All of the registrant's limited liability company member interests are directly owned by Duke Energy.
Duke Energy Ohio	All of the registrant's common stock is indirectly owned by Duke Energy.
Duke Energy Indiana	All of the registrant's common stock is indirectly owned by Duke Energy.

This combined Form 10-Q is filed separately by four registrants: Duke Energy, Duke Energy Carolinas, Duke Energy Ohio and Duke Energy Indiana (collectively the Duke Energy Registrants). Information contained herein relating to any individual registrant is filed by such registrant solely on its own behalf. Each registrant makes no representation as to information relating exclusively to the other registrants.

Duke Energy Carolinas, Duke Energy Ohio and Duke Energy Indiana meet the conditions set forth in General Instructions H(1)(a) and (b) of Form 10-Q and are therefore filing this form with the reduced disclosure format specified in General Instructions H(2) of Form 10-Q.

SELF ASSESSMENT PROBLEMS & QUESTIONS –CHAPTER 4

1. The 10-Q and 10-K Reports constitute mandatory reporting by publically traded firms, only.

   A. True
   B. False

2. The 8-K report contains less detailed information than the 10-K report and is filed on quarterly basis.

   A. True
   B. False

3. The purpose of an 8-K current report is to protect the interest of the investors.

   A. True
   B. False

4. Your firm has just been indicted for a serious breach of the EPA environmental regulations. This incident has a potential for tens of thousands of dollars in fines, and millions of dollars of compensation in civil law suits. Which report must be filed?

   A. 8-K report
   B. 10-Q report
   C. 10-K report
   D. 10-K and 8-K reports

5. If your firm is a sole proprietorship, privately traded, posts annual revenue to the order of $5 million and employs to (10) people your company is not required to file the 10-K, 10-Q and 8-K reports.

   A. True
   B. False

6. The SEC is a standards' agency and its recommendations constitute guidelines for the financial industry.

   A. True
   B. False

Chapter 5

Income Statement and Balance Sheet

INTRODUCTION

Financial statements, such as income statements and balance sheets, are not just an integral component of the federally mandated 10-K report, but they form the very foundation of our accounting system. Balance sheets and income statements serve as a financial "dashboard" for firms. Regardless of the reporting requirements, balance sheets and income statements are a vital management tool for all for-profit firms.

Comprehension of financial reports is important for energy engineers since most energy engineers are either employed by for-profit organizations or interface and transact with them, in some capacity. Balance sheets and income statements help energy engineers identify the following important parameters that are crucial to the development of a successful proposal:

1) Would a certain cost qualify as an expense or an asset?
2) Methods, steps and components associated with the computation of the net income or the bottom line.
3) Identification of financial parameters that are essential to the computation of financial ratios—and other financial indicators such as NPV and payback period—that gage and annunciate financial performance of firms, regardless of whether those firms are clients or service providers.

INCOME STATEMENT

Income statement is also referred to as the profit and loss statement. An income statement reports the amount of profits or loss incurred during a period of time, such as year ended December 31, 2009.

An income statement is based on the following basic accounting formula:

Pre-tax Income = Net Sales—Cost of Goods Sold
Net Income = Pre-tax Income—Income Tax

BALANCE SHEET

Balance sheet is also referred to as the statement of financial position. A balance sheet presents a picture of the firm's financial position as of a given point in time. A balance sheet is founded on the following fundamental accounting equation:

Total Assets = Total Liabilities + Total Shareholders Equity

Note that liabilities don't constitute assets; they, in combination with the stockholders investment in the firm, must equal the assets. For example, liabilities associated with accounts payable for raw materials are offset, on the other side of the equation, with value of inventory, under assets.

In order to study the composition and structure of income statements and balance sheets lets examine the financial statements of a firm, ABC Corporation, through Case Study 5-1.

CASE STUDY 5-1: ABC CORP. FINANCIAL STATEMENTS

This case study focuses on the financial statements of a publically held firm, ABC Corp. The financial statements, including the income statement and the balance sheet are appended below. However, in order to understand these statements clearly and to be able to utilize them for financial analysis accurately we must note the following contextual clarifications:

1) Most of the financial data for ABC Corp. is categorized in three columns, representing years 2003, 2004 and 2005. This 3-year compendium of financial data facilitates determination of the firm's financial performance trend.

2) These statements are intended to be explicit and allow the reader to see application and significance of most of the financial and accounting concepts presented in this text.

3) Some of the financial ratios, stated in the later segments of these statements, represent averages over the three years.

4) Typically, negative ratios are indicative of the fact that either the numerator or the denominator, in the pertaining ratio, is negative.

5) Certain expenses or losses, under certain circumstances, are reported as negative income. While this practice appears to be confusing and counterintuitive, it is not disallowed. An example of this is the statement of –$11.18 MM dollars on ABC's income statement, as "Non-Operating Income," under Column A3.

6) ABC Corp's income statement and the balance sheet emulate financial reports of a actual corporation listed on the NYSE, New York Stock Exchange. Most texts tend to draw simplistic examples for illustration purposes. However, in such cases, the simplistic models tend to filter out nuances and deviations from the mainstream, thus leaving the reader ill-equipped for deciphering the financial statements for many, real, publically held firms.

7) If you examine the financial reports, submitted by GE, Exxon, Pepsico, Verizon, or some other Fortune 100 firms, as a part of their 10-K reporting, you might notice that the format, layout and the terminology, on some financial statements, is somewhat different than the format or layout used for ABC. However, the parameters and financial data necessary for performance of financial analysis, i.e. computation of financial ratios, will be quite the same. While the figures, financial performance and trend may differ to a certain extent, the financial statements of most firms will tend to comport with ABC's statements in the following aspects:

   a. Format and content of the Income Statement, the Balance Sheet and the Cash Flow Statement.

   b. Statement of annual revenue, referred to as Net Sales, Cost of Goods Sold, Gross Profit, Expenses, Taxes, Net Income, Assets, Current Assets, sub-categories under Current Assets, Liabilities, Current Liabilities, sub-categories under Current Liabilities and Shareholders Equity.

   c. Financial Strength Ratios and Asset Ratios

ABC CORP.

Note: All figures are in $MM or millions of US dollars.

Column A1

Income Statement (SUMMARY), ABC Corp.	09/05	09/04	09/03
Net Sales	2,359.45	1,417.19	922.12
Cost of Goods Sold	1,475.59	811.11	570.58
Pretax Income	71.28	90.53	23.04
Net Income	46.83	55.78	15.48

Column A2

Balance Sheet (SUMMARY), ABC Corp.	09/05	09/04	09/03
Assets			
Total Current Assets	1,047.97	648.69	666.82
Net PP&E	304.32	182.40	150.61
Total Assets	**4,022.09**	1,634.15	1,545.29
Liabilities and Shareholders Equity			
Total Current Liabilities	557.38	396.84	397.01
Long-term Debt	2,252.04	784.25	848.82
Total Liabilities	**3,179.36**	1,316.73	1,343.29
Total Shareholders Equity	**842.73**	316.04	202.00
Total Liabilities & Shareholders Equity	**4,022.09**	1,634.15	1,545.29
Cash Flow Statement (SUMMARY)	**09/05**	**09/04**	**09/03**
Net Cash Flows from Operations	227.33	104.87	76.21
Net Cash Flows from Investing	-1,693.83	-68.58	-446.40
Net Cash Flows from Financing	1,481.86	-130.68	471.85

Detailed Annual Income Statement, ABC Corp.

Column A3

Period Ended	09/05	09/04	09/03
Net Sales	2,359.45	1,417.19	922.12
Cost of Goods Sold	1,475.59	811.11	570.58
Gross Profit	883.86	606.07	351.54

Expenditures:			
R&D Expenditure	**29.34**	23.19	14.36
Selling, General & Admin Expenses	**650.04**	426.66	277.54
Depreciation & Amortization	**n/a**	n/a	n/a
Non-Operating Income	**-11.18**	0.01	0.58
Interest Expense	**122.02**	65.70	37.18
Income Before Taxes	71.28	90.53	23.04
Prov. for Inc. Taxes	**24.45**	34.37	7.55
Minority Interest	**n/a**	n/a	n/a
Realized Investment (Gain/Loss)	**n/a**	n/a	n/a
Other Income	**n/a**	n/a	n/a
Net Income before Extra Items	**46.83**	56.16	15.48
Extra Items & Disc. Ops.	**n/a**	-0.38	n/a
Net Income	46.83	55.78	15.48

Detailed Balance Sheet, ABC Corp.

As of	**09/05**	**09/04**	**09/03**

Column A4

Assets			
Cash 29.85	13.97	107.77	
Marketable Securities	n/a	n/a	n/a
Receivables	373.40	289.63	270.58
Total Inventories	**451.55**	**264.73**	**219.25**
Raw Materials	117.18	47.88	60.73
Work In Progress	37.93	31.38	34.91
Finished Goods	296.45	185.46	123.61
Notes Receivable	n/a	n/a	n/a
Other Current Assets	193.17	80.37	69.21
Total Current Assets	**1,047.97**	648.69	666.82
Property, Plant & Equipment, Net	304.32	182.40	150.61
Property, Plant & Equipment, Gross	505.75	355.36	301.43
Accumulated Depreciation	201.42	172.96	150.82

Interest and Advance to Subsidiaries	n/a	n/a	n/a
Other Non-Current Assets	n/a	n/a	n/a
Deferred Charges	89.58	62.67	79.02
Intangibles	2,580.22	740.40	648.85
Deposits & Other Assets	n/a	n/a	n/a
Total Assets	**4,022.09**	1,634.15	1,545.29
Liabilities			
Notes Payable	n/a	n/a	n/a
Accounts Payable	281.95	226.23	172.63
Current Long-Term Debt	37.90	22.12	70.47
Current Port. Cap Lease	1.41	1.77	2.38
Accrued Expense	195.65	125.04	130.96
Income Taxes	**40.47**	21.67	20.57
Other Current Liabilities	n/a	n/a	n/a
Total Current Liabilities	**557.38**	396.84	397.01
Mortgages	n/a	n/a	n/a
Deferred Charges/Inc.	208.25	7.27	n/a
Convertible Debt	n/a	n/a	n/a
Long-Term Debt	2,252.04	784.25	848.82
Non-Curr. Capital Leases	15.98	21.75	21.72
Other Long-Term Liab.	145.71	106.61	75.73
Total Liabilities	3,179.36	1,316.73	1,343.29
Shareholder Equity			
Minority Interest	n/a	1.38	n/a
Preferred Stock	n/a	n/a	n/a
Common Stock	0.67	0.64	0.62
Capital Surplus	671.38	224.96	185.56
Retained Earnings	267.32	220.48	164.70
Treasury Stock	70.82	130.07	130.07
Other Liabilities	-25.81	0.03	-18.81
Total Shareholders Equity	842.73	316.04	202.00
Total Liabilities & Shareholders Equity	**4,022.09**	1,634.15	1,545.29

I. NET INCOME CALCULATION

The net income calculation can be demonstrated by the application of the net income formula to **Column A3** in the detailed version of the income statement, as follows:

Net Income = (Net Sales – Cost of Goods Sold) – Expenditures – Taxes

Net Income = (2,359.45 – 1,475.59) – (29.34 + 650.04 + 11.18 + 122.02) – 24.45 = $46.83 million

Note that the non-operating income for 2005 is listed as –11.18. This can cause some ambiguity in the treatment of this line item when calculating the net income. One could be tempted to actually treat this entry as a negative number in the summation of all of the expenses for 2005. This would yield an incorrect net income because the addition of –11.18 would have the net effect of reducing the overall expense. The –$11.18 MM entry for non-operating Income, in fact, implies that ABC experienced a loss of $11.18 MM, in 2005, in the non-operating income category.

II. BALANCE SHEET CALCULATION

The use of the balance sheet calculation formula can be illustrated by application of this formula to column A3 in the income statement below.

Total Assets = Total Liabilities + Total Shareholders Equity
= 3,179.36 + 842.73
= 4,022.09 or $4,022 million

III. TOTAL ASSET CALCULATION ON ASSET SIDE OF THE BALANCE SHEET AND THE ROLE OF DEPRECIATION:

Total Assets = Current Assets + Net Value of Plant & Equipment + Deferred Charges + Intangible Assets

Or, in expanded form:

Total Assets = Current Assets + (Gross Value of Plant & Equipment – Accumulated Depreciation) + Deferred Charges + Intangible Assets

For ABC's fiscal year 2005:

Total Assets = 1047.9 + (505.75 – 201.42) + 89.58 + 2580.22
= 4,022.09 or $4,022 million

Note that "Property, Plant & Equipment, Net," "Property, Plant & Equipment, Gross," and "Accumulated Depreciation" are all listed on ABC's balance sheet, simultaneously. This could mislead one to add all three of these figures to the total current assets in the derivation of the total assets. Also note that accumulated depreciation is stated as a positive number, again, somewhat confusing. As illustrated through the calculation above, the accumulated depreciation must be treated as a negative number in determination of the net value of plant and equipment from the gross value of plant and equipment. This leads us to an important observation: the financial analysis using the financial statements such as Balance Sheets and Income Statements requires more than pure arithmetic; it requires knowledge and understanding of the parameters and associated concepts. In this case, we see that recognition of the fact that accumulated depreciation reduces gross assets to yield net assets is crucial in accurate assessment of total assets.

End-of-chapter problems and questions are designed to facilitate greater understanding of financial statements and typical entries therein. Most of these problems are based on the ABC Case Study 5-1. These problems expand the study of balance statements and income statements to include approximate product price determination, especially for commodity products, net income trend, etc.

FINANCIAL STATEMENTS EXAMPLE IN THE ENERGY INDUSTRY

Even though ABC Corp. case study pertains to a non-energy firm, the conventions, format, typical financial entries and the sorting of financial data into proper segments/fields within the financial statements, apply to firms in the energy industry just as well. Nevertheless, in order to broaden our perspective on the subject of financial statements we will examine a recent submission of "Condensed Consolidated Statements of Operations" by Duke Energy.

As we examine the financial statements for Duke Energy, below, we notice some of the following distinct differences as compared to the ABC financial statements:

1) The Duke Energy financial quarterly (10-Q) report, ending March, 31, consists of the following three financial statements:

   a) Income statement for 2010
   b) Balance sheet for 2010
   c) Cash flow statement for 2010

The first section of the income statement breaks out the total first quarter, 2010, revenue of $3,594 million for the Duke's three business segments. This revenue distribution between the three business segments is shown in Figure 5-1. This portion of Duke Energy income statement provides its investors an immediate perspective on the composition of the firm. It shows, for example, that Regulated Electric is Duke Energy's mainstay. And, even though Duke has diversified into natural gas, this segment of the business remains a small piece of the pie. The income statement for ABC Corp., in Case Study 5-1, only contained the total revenue for the firm, and does not provide enough details to the investor on the operational make up or revenue distribution of the company.

2) Since Duke is in the business of generating, transmitting and selling electrical energy, and distribution/sale of natural gas and some coal, the cost side of its business is dominated by fuel used in its operations. Fuels constitute 45% of the total operating expenses. As we compare

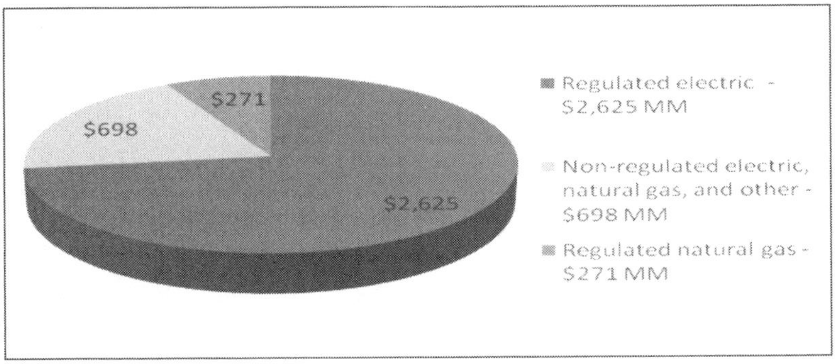

Figure 5-1. Duke Energy Revenues, First Quarter, 2010

Duke's expenses with ABC's we notice that ABC, being a non-energy consumer product manufacturer, has over 60% of its cost categorized as cost of goods sold, for 2005.

3) Overall, Duke Energy reports contain more details in most categories of financial statements, as compared to ABC.

4) Duke's income statement concludes with significant amount of data on its stocks; i.e. EPS, earnings per share, dividends per share, number of shares outstanding, etc.

End-of-chapter problems and questions pertaining to Duke Energy example allow an opportunity to hone ones understanding of energy industry related financial statements and typical entries therein.

DUKE ENERGY CORPORATION

CONDENSED CONSOLIDATED STATEMENTS OF OPERATIONS[1]

(Unaudited)

(In millions, except per-share amounts)

Item 1. Financial Statements

Income Statement	**Three Months Ended March 31,**	
	2010	**2009**
Operating Revenues		
Regulated electric	$ 2,625	$2,545
Non-regulated electric, natural gas, and other	698	467
Regulated natural gas	271	300
Total operating revenues	3,594	3,312
Operating Expenses		
Fuel used in electric generation and purchased power - regulated	819	849
Fuel used in electric generation and purchased power - non-regulated	278	148
Cost of natural gas and coal sold	190	222
Operation, maintenance and other	899	811
Depreciation and amortization	456	414
Property and other taxes	193	193
Total operating expenses	2,835	2,637

Gains on Sales of Other Assets and Other, net	2	6
Operating Income	**761**	**681**
Other Income and Expenses		
Equity in earnings of unconsolidated affiliates	29	6
Other income and expenses, net	91	22
Total other income and expenses	120	28
Interest Expense	210	184
Income From Continuing Operations Before Income Taxes	671	525
Income Tax Expense from Continuing Operations	226	179
Income From Continuing Operations	445	346
Income From Discontinued Operations, net of tax	—	3
Net Income	**445**	**349**
Less: Net Income Attributable to Noncontrolling Interests	—	5
Net Income Attributable to Duke Energy Corporation	$445	$344
Earnings Per Share - Basic and Diluted		
Income from continuing operations attributable to Duke Energy Corporation common shareholders		
Basic	$0.34	$0.27
Diluted	$0.34	$0.27
Income from discontinued operations attributable to Duke Energy Corporation common shareholders		
Basic	$—	$—
Diluted	$—	$—
Net income attributable to Duke Energy Corporation common shareholders		
Basic	$0.34	$0.27
Diluted	$0.34	$0.27
Dividends per share	$0.24	$0.23
Weighted-average shares outstanding		
Basic	1,310	1,282
Diluted	1,311	1,283

See Notes to Unaudited Condensed Consolidated Financial Statements

Table of Contents
PART I

DUKE ENERGY CORPORATION
CONDENSED CONSOLIDATED BALANCE SHEETS
(Unaudited)
(In millions)

	March 31 2010	**December 31 2009**
ASSETS		
Current Assets		
Cash and cash equivalents	$1,080	$1,542
Receivables (net of allowance for doubtful accounts of $45 at March 31, 2010 and $42 at December 31, 2009)	724	845
Restricted receivables of variable interest entities (net of allowance for doubtful accounts of $32 at March 31, 2010 and $6 at December 31, 2009)	1,200	896
Inventory	1,337	1,515
Other	1,150	968
Total current assets	5,491	5,766
Investments and Other Assets		
Investments in equity method unconsolidated affiliates	363	436
Nuclear decommissioning trust funds	1,846	1,765
Goodwill 4,349	4,350	
Intangibles, net	571	593
Notes receivable	127	130
Other	2,661	2,533
Total investments and other assets	9,917	9,807
Property, Plant and Equipment		
Cost 56,493 55,362		
Less accumulated depreciation and amortization	17,787	17,412
Net property, plant and equipment	38,706	37,950
Regulatory Assets and Deferred Debits		
Deferred debt expense	254	258
Regulatory assets related to income taxes	618	557
Other	2,647	2,702
Total regulatory assets and deferred debits	3,519	3,517
Total Assets $57,633	**$57,040**	

See Notes to Unaudited Condensed Consolidated Financial Statements

Table of Contents

PART I

DUKE ENERGY CORPORATION

CONDENSED CONSOLIDATED BALANCE SHEETS—(Continued)

(Unaudited)

(In millions, except per-share amounts)

	March 31 2010	December 31 2009
LIABILITIES AND EQUITY		
Current Liabilities		
Accounts payable	$1,299	$1,390
Notes payable	12	—
Non-recourse notes payable of variable interest entities	350	—
Taxes accrued	383	428
Interest accrued	251	222
Current maturities of long-term debt	586	902
Other	973	1,146
Total current liabilities	3,854	4,088
Long-term Debt	15,900	15,732
Non-recourse long-term debt of variable interest entities	379	381
Deferred Credits and Other Liabilities		
Deferred income taxes	6,040	5,615
Investment tax credits	337	310
Asset retirement obligations	3,238	3,185
Other 5,837	5,843	
Total deferred credits and other liabilities	**15,452**	**14,953**
Commitments and Contingencies		
Equity		
Common Stock, $0.001 par value, 2 billion shares authorized; 1,312 million and 1,309 million shares outstanding at March 31, 2010 and December 31, 2009, respectively	1	1
Additional paid-in capital	20,697	20,661
Retained earnings	1,589	1,460
Accumulated other comprehensive loss	(373)	(372)
Total Duke Energy Corporation shareholders' equity	21,914	21,750
Noncontrolling interests	134	136
Total equity	**22,048**	**21,886**
Total Liabilities and Equity	**$57,633**	**$57,040**

See Notes to Unaudited Condensed Consolidated Financial Statements

Table of Contents
PART I

DUKE ENERGY CORPORATION
CONDENSED CONSOLIDATED STATEMENTS OF CASH FLOWS
(Unaudited)
(In millions)

	Three Months Ended March 31	
	2010	**2009**
CASH FLOWS FROM OPERATING ACTIVITIES		
Net Income	$445	$349
Adjustments to reconcile net income to net cash provided by operating activities:		
Depreciation and amortization (including amortization of nuclear fuel)	506	463
Equity component of AFUDC	(55)	(28)
Severance expense	68	—
Gains on sales of other assets	(1)	(11)
Deferred income taxes	162	165
Equity in earnings of unconsolidated affiliates	(29)	(6)
Contributions to qualified pension plans	—	(500)
(Increase) decrease in		
Net realized and unrealized mark-to-market and hedging transactions	3	(23)
Receivables	94	222
Inventory	180	(110)
Other current assets	14	24
Increase (decrease) in		
Accounts payable	(114)	(244)
Taxes accrued	(52)	(26)
Other current liabilities	(179)	(176)
Other assets	81	73
Other liabilities	(2)	18
Net cash provided by operating activities	1,121	190
CASH FLOWS FROM INVESTING ACTIVITIES		
Capital expenditures	(1,179)	(845)
Investment expenditures	(20)	(61)
Purchases of available-for-sale securities	(591)	(930)
Proceeds from sales and maturities of available-for-sale securities	550	917
Net proceeds from the sales of other assets, and sales of and collections on notes receivable	3	38
Purchases of emission allowances	(5)	(25)
Sales of emission allowances	7	15
Change in restricted cash	1	3

Other	(2)	(6)
Net cash used in investing activities	(1,236)	(894)
CASH FLOWS FROM FINANCING ACTIVITIES		
Proceeds from the:		
Issuance of long-term debt	451	1,916
Issuance of common stock related to employee benefit plaPayments for the redemption of long-term debt (609)	(603)	
Notes payable and commercial paper	93	(263)
Dividends paid	(316)	(296)
Other 3 (6)		
Net cash (used in) provided by financing activities	(347)	919
Net (decrease) increase in cash and cash equivalents	(462)	215
Cash and cash equivalents at beginning of period	1,542	986
Cash and cash equivalents at end of period	**$1,080**	**$1,201**
Supplemental Disclosures:		
Significant non-cash transactions:		
Debt associated with the consolidation of Cinergy Receivables	$257	$—
Accrued capital expenditures	$473	$320

[1]See complete financial statements and Duke Energy and its business segments, in conjunction with pertinent notes to Unaudited Condensed Consolidated Financial Statements, on Duke Energy web page at: http://finance.yahoo.com/q?s=duk

SELF ASSESSMENT PROBLEMS & QUESTIONS—CHAPTER 5

1. An Income Statement is founded on the equation listed below:
   Total Net Income = Net Sales/Revenue – COGS – Income Tax

   A. True
   B. False

2. Which financial statement or report is likely to contain a direct reference to the accumulated annual depreciation?

   A. Annual Cash Flow Statement
   B. Annual Balance Sheet
   C. 10-Q Report
   D. Annual Income Statement

3. In Case Study 5-1, financial statements show that ABC's net income rose, from 2003 to 2005, by:

   A. 33%
   B. 150%
   C. 303%

4. Examine the ABC Corp. financial statements and the financial data provided under Column A2. Add liabilities and equity for 2004. The sum, in accordance with the balance sheet equation is:

   A. $1,634.15 MM
   B. $185.56 MM
   C. $4,022.09 MM

5. Retained earnings for ABC Corp., for Year 2003 are:

   A. $164,700,000
   B. $267,320,000
   C. $220.48 MM

6. The income tax bracket for ABC Corp., based on 2004 data, is

   A. 37.97%
   B. 34.8%
   C. 42.13%.

7. ABC Corp. has long term debt associated, most likely, with plant and equipment.

   A. True
   B. False

8. Assume that the sales volume for ABC Corp., in 2003, was 100MM units and that ABC produced one commodity product, what was the price per unit?

   A. $41.59
   B. $9.22
   C. $36.50
   D. $14.17

9. What is the difference in size between ABC and Duke Energy, based on revenue information provided in the most recent financial statements?

   A. Duke is approximately 50% larger
   B. ABC is approximately 40% larger
   C. Duke is approximately 30% larger

10. Which firm, ABC or Duke, shows greater growth based on revenue over two (2) consecutive years?

   A. Duke shows greater growth.
   B. ABC shows greater growth.

11. Which firm, ABC or Duke, registered greater growth based on net income over the most recent consecutive years?

   A. Duke shows greater growth.
   B. ABC shows greater growth.
   C. ABC shows a decline in net income.
   D. Both A and C

12. Depreciation of property, plant and equipment recorded by Duke in March 2010 was:

   A. $17,412 MM
   B. $17,787 MM
   C. $55,362 MM
   D. $38,706 MM

Chapter 6

Financial Metrics and Ratios

INTRODUCTION

Now that we have an appreciation of financial statements, let's examine some financial tools and techniques that facilitate assessment of the financial standing of firms. Some of these tools are simple financial ratios, while others require multistep calculations.

Understanding of financial ratios and metrics is important for business leaders, program managers, project managers, production managers, energy engineers and energy managers. Financial ratios and other financial metrics allow managers and engineers to take the "pulse" of projects, programs, facilities and businesses. Regardless of how state of the art, avant-garde, attractive and appealing a proposal, program or project is—from qualitative perspective—ultimately, it must yield acceptable and competitive financial returns. Financial ratios and financial metrics allow us to do just that, objectively and quantitatively.

If an energy engineer is interested in knowing whether his or her project will measure up to the reining corporate thresholds, expectations and standards of their clients or employers, they need to be aware of the mechanics and composition of pertinent financial metrics.

While certain firms and institutions assign more weight to one financial metric or ratio versus another, in this chapter we will focus on following more mainstream ones:

1) NPV
2) Payback Period
3) ROI
4) ROR
5) ROE
6) IRR
7) Working Capital
8) Inventory Turnover Ratio
9) Debt to Equity Ratio

NET PRESENT VALUE, NPV

The concept of NPV applies to series of cash flows. NPV is the sum of present values of cash flows.

$$NPV = CF_1/(1 + i)^1 + CF_2/(1 + i)^2 + CF_3/(1 + i)^3 + \ldots + CF_n/(1 + i)_n$$

Where,

CF_1 = Cash Flow at the end of 1st year
CF_2 = Cash Flow at the end of 2nd year
CF_n = Cash Flow at the end of nth year
i = Interest Rate, in decimal format
n = Number of Years or Compounding Periods

PAYBACK PERIOD

Payback period, in terms of years, months or weeks, is simply the amount of time it takes for an investment to be recouped.

Example 6-1

A firm invests $40,000 in replacement of an old motor with a new premium efficiency motor. This projects results in net savings of $10,000 per annum, through energy savings, maintenance cost reduction and production loss avoidance. Find simple payback period for this investment.

Payback Period = Investment/Net Savings per year
= $40,000/$10,000 per year
= 4 Years

RETURN ON INVESTMENT, ROI

The return on investment ratio, ROI, compares net income and the average total assets of the firm. The purpose of this ratio is to measure the profit earned per dollar of investment.

ROI = ROC (Return On Capital) =
Net Income/Average Total Assets or Investment

Example 6-2

Calculate the payback period for the energy efficient motor project in Example 6-1.

ROI = ROC = \$10,000/\$40,000 = 0.25 or 25%

RATE OF RETURN, ROR

Rate of return is defined as the effective annual interest rate at which an investment accrues income. In a project investment situation, ROR would represent an interest rate that would yield equivalent profits if all of the money was, instead, invested at that rate.

The method for calculating ROR is iterative, and therefore, somewhat tedious. It requires calculation of present value at a few assumed and reasonable discount rates. Then interpolation and graphical straight line analysis are performed to calculate a more accurate value of ROR.

RETURN ON EQUITY, ROE

Return on equity is one of the profitability ratios. Profitability ratios are measures of performance that indicate the amount that firm is earning relative to some base, such as sales, assets, or equity.

ROE gauges management's performance with respect to the interest of the stockholders. Although ROI, or return on assets, give an aggregate measure of the firm's performance, it does not gauge how the management is performing for the stockholders.

ROE = Earnings After Taxes/Equity

Where, equity = sum of common stock, additional paid-in capital (if any) and retained earnings.

INTERNAL RATE OF RETURN, IRR

Internal rate of return is an average of *all* cash flows, or net savings generated by the project, during its 10-year life expectancy, divided by the total investment. For illustration, see Case Study 6-1.

Application of financial metrics and ratios, applicable to projects, programs, measures and investments—in the energy and non-energy arena—is illustrated through Case Study 6-1.

CASE STUDY 6-1: ENERGY PROJECT—EQUIPMENT REPLACEMENT:

This case study is premised on an investment to replace a piece of equipment, at a capitalized cost of $200,000. All data and financial analysis pertinent to this case study are captured in Spreadsheet 6-1. It is assumed that the project is commissioned into service—and meets the capital justification criteria—on the first day of the first year of commissioning. It is further assumed that the existing equipment that forms the basis of this case study is obsolete, energy inefficient and is exhibiting short MTBF, mean time between failure. Each maintenance incident, planned or unplanned, results in loss of production, and in some cases, loss of market share. This improvement or benefit is accounted for under the productivity improvement row in the spreadsheet.

As we delve into this case study, we need to recognize the following additional points and clarifications:

- The distinction between initial cost, operating costs and the overall life cycle cost.
  - —The initial cost of $200,000 includes not only the equipment cost but the cost of installing, testing and commissioning the equipment or system.
  - —The operating cost includes of the cost of electricity for operating this equipment/system, at $32,675 per year.
  - —Since the cost segment of financial analysis must include all costs, over the life of the system, costs such as maintenance and consumable supplies must be taken into account, as well. Since costs constitute negative cash flows, they are shown in parenthesis, as negative numbers.

- Note that in this case study, cost reductions, by virtue of this investment, in form of energy cost savings, productivity improvement, maintenance cost reduction and safety incident cost aversion, are all savings that are considered as "earnings," realized through this investment. Therefore, all savings are shown as positive cash flows.

- The financial analysis conducted in this case study span 10 years which is the expected or assumed life span of the system or measure. The costs or earnings pertaining to each of the 10 years are tabulated as cash flows for the respective years. All negative cash flows (costs) and positive cash flows (savings or earnings) are summed for each year to represent the total or net cash flow for the year. The cash flows for each year are then converted into their present values. These present values of annual cash flows are summed to obtain the Net present value, or NPV, for the investment. Alternatively, this NPV calculation can be performed within a Microsoft Excel ® spreadsheet by embedding the NPV calculation function. NPV comparison between multiple investment alternatives is an objective and established method for arriving at financial analysis based investment decisions.

  When using the NPV metric for decisions on investments, the following "rules of thumb" apply:

  —An investment that yields a negative NPV constitutes a investment that yields a financial loss and should be rejected.

  —When comparing multiple alternatives, all offering positive NPVs, select the alternative with highest positive NPV.

  —When comparing alternatives involving net cost or net negative cash flow, select the alternative with lowest negative NPV.

- The financial analyses in this spreadsheet are premised on an assumed prevailing interest rate, or discount rate, of 10%. However, for internal projects or programs, within a firm, it is common for firms to apply an interest rate established by the corporation.

- Significant MSD, musculoskeletal disorder type injuries, associated with sprains and strains, can result in medical costs ranging from

$10,000 to $25,000, with litigation and workman's compensation settlements escalating the overall cost by tens of thousands of dollars. This is an "order of magnitude" approximation for costs associated with some common forms of injuries in industrial or commercial environment. When historical or statistical safety incident cost and worker compensation data is available, and if the new system or investment delivers a safer work environment, it is recommended that injury cost avoidance be claimed as a saving or earnings attributable to the project. In Case Study 6-1, based on historical data available and the past injury statistics in the current operation, it is assumed that a minimum of $10,000 will be saved, annually, due to the ergonomic design of the new equipment.

- The approach and format used for financial analysis in Case Study 6-1 is commonly used by firms for determining the financial merit of a project, proposal or measure. As shown in chapter 11, this approach can be adapted for different types of projects, including energy projects involving EPC or ESCOs.

- Comprehensive results, for the entire 10-year project life are shown in Appendix B. Cash flows for all 10 years can be seen in the comprehensive version.

Some of the ratios and metrics that are more commonly utilized in management of facilities—in the energy or non-energy realm—are addressed in the sections to follow.

WORKING CAPITAL

Working capital is a financial metric which represents operating liquidity or cash available to a business. Along with fixed assets such as plant and equipment, working capital is considered a part of operating capital. The formula for calculating the working capital is as follows:

Working Capital = Current Assets – Current Liabilities

Working capital is considered to be difference between current assets

Spreadsheet 6-1—Case Study 6-1

Description	Year 1	Year 3	Year 5	Year 7	Year 10
	In $1,000s				
Costs					
Write-off of Existing Investment,	($10)				
Total Initial Cost of Project	($200)				
Savings	In $1,000s				
Utility & Energy Savings - 32%	$10.456	$10.456	$10.456	$10.456	$10.456
Productivity Improvement - 35%	$11.436	$11.436	$11.436	$11.436	$11.436
Annual Maintenance Cost Savings	$30.00	$30.00	$30.00	$10.00	$10.00
Savings—Injury Avoidance	$10.00	$10.00	$10.00	$10.00	$10.00
Net Annual Cash Flows	($148.11)	$61.892	$61.892	$51.892	$51.892
Present Values of Annual Cash Flows	($134.64)	$46.50	$38.430	$26.629	$20.007

Financial Analyses Results, Actual Values	
NPV	$161,188
IRR	17%
ROI	31%
Payback Period	3.62 Years

and current liabilities. If current assets are less than current liabilities, an entity has a **working capital deficiency**, also called a **working capital deficit**. An extension or derivative of working capital is a term called 'net working capital." The formula below defines net working capital, mathematically:

Net Working Capital = Working Capital – Cash (A Current Asset) – Interest Bearing Liabilities (Short Term Liabilities)

CURRENT RATIO

Current ratio compares a firm's current assets and current liabilities.

Current Ratio = Current Assets/Current Liabilities

In this formula, current assets include cash, accounts receivable, inventories and prepaid expenses. Current liabilities include accounts payable, notes payable and taxes payable.

Example 6-3

Calculate current ratio for the ABC Corp. for year 2003, in Case Study 5-1, Chapter 5.

Current Ratio = Current Assets/Current Liabilities
= $666.82 MM/$397.01 MM
= 1.7

ACID TEST RATIO

This ratio compares the firm's most liquid assets and liabilities.

Acid Test Ratio = (Current assets – Inventories)/Current Liabilities

Where, current assets include cash, accounts receivable, inventories and prepaid expenses. Current liabilities include accounts payable, notes payable and taxes payable.

Example 6-4

Calculate the Asset Test Ratio for ABC Corporation, year 2003.

Acid Test Ratio = (Current assets – Inventories)/Current Liabilities
Acid Test Ratio = (666.8 – 219.25)/397 = 1.127

PLANT TURNOVER RATIO

This ratio compares a firm's sales revenue with the dollar investment in plant and equipment; that is required to generate the revenue.

Plant-turnover Ratio = Sales Revenue/Investment

Example 6-5

Calculate the plant turnover ratio for a corporation with annual sales revenue of $2,000,000 and investment in equipment and plant of $500,000:

Plant Turnover Ratio = $2,000,000/$500,000 = 5 times

INVENTORY TURNOVER RATIO

Inventory turnover ratio is defined as the ratio of cost of goods sold to average inventory.

The formula for inventory turnover ration is:

Inventory-turnover Ratio = Cost of Goods Sold/Average Inventory

The formula for average inventory is

Average Inventory = (Initial Inventory + Final Inventory)/2

A low turnover ratio could be indicative of the overstocking, perhaps, in anticipation of higher prices, eminent strikes, material shortage, etc. However, low inventory turnover ratio could result in high WIP, work in process, lower plant ROI, fiscal inefficiency and the risk of obsolescence. A high turnover ratio could be indicative of inadequate inventory levels, exposing business to production interruptions and possible loss of business.

Example 6-6

Calculate the inventory turnover ratio for ABC Corp., in Case Study 5-1, for year 2005.

Inventory Turnover Ratio = Cost of Goods Sold/Average Inventory
= $1,475.59 MM/$451.55 MM = 3.27

DEBT TO EQUITY RATIO

Debt to equity ratio, sometimes denoted by D/E, is a firm's or shareholder's debt to the shareholder's equity. The debt and equity used in the computation of this ratio are typically retrieved from the firm's balance sheet. This ratio may also be calculated using market values of firm's debt and equity if the firm is publically traded. One common source for market value of publically held firms' debt and equity is the Standards and Poor's or S&P.

The formula for debt to equity ratio is:

D/E = Debt (or Liabilities)/Equity

Example 6-7

Calculate the debt to equity ratio for Duke Energy, based on the quarterly, March 31, 2010, financial statements exhibited under Example 5-1, Chapter 5. Note: Unless otherwise specified, use the totals data available for both debt and equity.

D/E = Debt (or Liabilities)/Equity
= $15,452 MM/$22,048 MM
= 0.701 or 70.1%

SELF ASSESSMENT PROBLEMS & QUESTIONS—CHAPTER 6

1. The concept of time value of money would not apply to economies and countries in the world where practice of charging interest for loans or

disbursement of interest for securities is outlawed.

A. Yes
B. No

2. Net present value of an investment would rise as the interest rate rises.

A. True
B. False

3. Inclusion of safety related cost savings in financial investment analysis tends to increase expenses and lower the NPV.

A. True
B. False

4. Annual energy cost for operating the new, energy efficient, equipment in Case Study 6-1 is:

A. $32,675
B. $27,774
C. $4,901
D. $22,219

5. Assume that a plant turnover ratio is the same as an asset turnover ratio, calculate the asset turnover ratio for ABC Corp. for 2005.

A. 0.72
B. 0.87
C. 0.6.
D. 0.59

6. Calculate the inventory turnover ratio for ABC Corp., in Case Study 5-1, for year 2004.

A. 3.72
B. 0.87
C. 2.6
D. 3.06

7. What was the working capital for Duke Energy in the first quarter, 2010? Use the March 31, 2009, financial statements exhibited under

Example 5-1, Chapter 5.

A. $251,850,000
B. $1,557,380,000
C. $1,637,000,000
D. $2,617,000,000

8. Calculate the debt to equity ratio for Duke Energy, based on the quarterly, March 31, 2009, financial statements exhibited under Example 5-1, Chapter 5.

A. 54.3%
B. 68.3%
C. 71.2%
D. 46.52%

Chapter 7

Depreciation Alternatives

INTRODUCTION

Depreciation plays an important role in accounting for the attrition in the useful life of assets. As such, periodic depreciation claimed by a firm offsets the book value of the assets by en equivalent amount. Depreciation claimed by firms, such as the $414 MM depreciation recorded in Duke Energy's March 31, 2009, quarterly financial statements, in Chapter 5, is treated as an expense. Therefore, depreciation has a direct impact on the bottom line, or the net income.

While, typically, program managers, project managers, production managers, energy engineers and energy managers are not directly influenced by depreciation—nor do they, in most cases make decisions that have direct ramifications on depreciation—understanding depreciation and the various methods of depreciation is important. At times, federal or state governments do permit accelerated depreciation of certain assets in order to incentivize and expand their application and acceptance. Such incentives, for example, may be offered for renewable energy alternatives in an effort to stimulate installation and use of such measures. Accelerated depreciation results in greater annual depreciation during the early years of asset operation; in essence, reducing the gross income, or the income before taxes. Since tax is premised on the gross income, accelerated depreciation, in effect, reduces the tax liability of the firm. Therefore, energy engineers, who are cognizant of depreciation and its financial ramifications in the energy arena, would tend to be better equipped to parlay the tax advantages associated with depreciation in making their proposals financially attractive.

There are several methods of depreciation. However, in this text, we will devote attention to the ones that are mainstream and more applicable to energy or non-energy projects in the industrial and non-industrial domains. Prior to discussing depreciation alternatives, we need to clarify certain finance and accounting terms that constitute the depreciation computation methods.

DEPRECIATION BASIS OF AN ASSET

Depreciation basis of an asset is defined, mathematically, as:

$$\text{Depreciation Basis} = \text{Purchase Price} - S_n$$

Where

S_n = Salvage value or book value at the end of the service life of the capital asset.

"n" = represents the life span of the asset, in years.

PURCHASE PRICE OR TOTAL INITIAL COST:

Purchase price or total initial cost of a project is an all-inclusive cost, that in most cases would include the design cost, engineering cost, equipment cost, real estate if needed, necessary infrastructure, installation cost, start-up cost, commissioning/certification cost. Often, such all-inclusive cost comprises the turnkey cost of a project.

BOOK VALUE:

Book value is, essentially, the "un-depreciated" value of an asset at given point in life of the asset. Book value of an asset can, mathematically, be defined as follows:

Book Value of an investment or asset, after "n" years of depreciation = Total Initial Investment – Total Depreciation taken in "n" years.

STRAIGHT LINE (S/L) METHOD

$$\text{Depreciation} = D = (C - S_n)/n$$

Where

D = Annual depreciation

C = Initial cost or purchase price

S_n = Salvage value or book value at the end of the service life of the capital asset.

n = Life span of the asset, in years.

SUM-OF-THE-YEARS' DIGIT (SOYD) METHOD:

$$Dj = \frac{(C - S_n)(n - j + 1)}{0.5\, n\, (n + 1)}$$

Where

Dj = Depreciation in year "j"
C = Initial cost or purchase price
S_n = Salvage value or book value at the end of the service life of the capital asset.
n = Life span of the asset, in years.

DOUBLE DECLINING BALANCE METHOD

This method is independent of the salvage value. This method depends on accumulated depreciation.

$$D_j = 2(C - \sum_{m=1}^{j-1} D_m)$$

Where

D_j = Depreciation in year "j"
C = Initial cost or purchase price
D_m = Depreciation in the middle of the asset's life.
m = Middle of asset's life span

STATUTORY DEPRECIATION SYSTEMS

Statutory depreciation systems include the accelerated cost recovery system (ACRS) and the modified accelerated cost recovery system (MACRS).

The ACRS applies to property placed in service in 1981 and thereafter.

The MACRS applies to property placed in service after 1986. Other methods i.e. S/L, DDB, etc. cannot be used except in special cases.

D_j = C x Factor

Where

D_j = Depreciation in the "j^{th}" year
C = Initial cost or purchase price

The factor is identified through tax publications and depends on the asset's cost recovery period. For instance, factors for the 3rd year, in a 5-year recovery period, are 0.21 under ACRS and 0.1920 under MACRS.

DEPRECIATION METHOD SELECTION

Depreciation method selection is premised on how fast the business wishes to depreciate the asset, to its salvage value. If a business prefers to pursue accelerated or "front loaded" depreciation, methods such as the double declining balance method should be considered. However, if steady depreciation comports well with the firm's overall business strategy, methods such as the straight line depreciation method would be more suitable.

SELF ASSESSMENT PROBLEMS & QUESTIONS—CHAPTER 7

1. As a manager and owner of a firm you are interested in minimizing your firm's income tax for the current year. Which of the following two depreciation methods should you avoid?

   A. Double declining balance method
   B. Straight line method

2. You have just expanded your manufacturing operations at a total cost of $2,000,000. What would the first year's depreciation expense be if you are to use the ACRS method and the allowed "factor" for the first year is 0.2?

   A. $200,000
   B. $140,000
   C. $700,000
   D. $400,000

3. If the depreciation expense in Duke Energy's March 31, 2010, quarterly financial statements is reduced by 50%, what would be the revised operating income?

   A. $989 MM
   B. $509 MM
   C. $305 MM

4. The 50% reduction of depreciation in problem 3 will:

   A. Increase Duke Energy's tax liability
   B. Increase net income
   C. Decrease Duke's tax liability
   D. Cause A & B

5. MACRS offers less annual depreciation as compared with ACRS.

   A. True
   B. False

6. Corporations have the latitude to chose any depreciation method that suits their fiscal objectives for a given year.

   A. True
   B. False

7. Depreciation period is always linked to the life expectancy of a project or asset.

   A. True
   B. False

Chapter 8

Inventory Concepts

INTRODUCTION

The general definition of inventory is that it is a stock of items that are kept on hand by firms to meet their customers' demands. In a manufacturing facility, all raw materials, components, pre-fabricated parts, and sub-assemblies are categorized as incoming goods or un-finished goods inventory. Some firms, with processes that are energy or utility intensive, extend as far as including electricity, gas, fuel oil and water in the raw materials category. Of course, for energy firms in the business of generating electric power from non-renewable energy sources, such as, fossil fuel, natural gas, coal and nuclear, all of the preceding types of fuels would constitute raw materials and, thus, be accounted for as incoming goods inventory.

Once products are manufactured, processed and finished, they get categorized as finished goods inventory. In the electric power generation business the finished good is electricity. Unlike tangible storable products like automobiles, electronic devices and food products, electricity is not easily stored, except in form of energy through systems like pump storage systems. Therefore, electricity, unlike fossil fuels, is difficult to "inventory" in the traditional sense. However, in the electricity domain, one can ***inventory*** the capacity in terms of kW, MW, GW, etc. Of course, usage of this capacity would have to be considered in form of energy, measured in kWh.

While inventory is an innate component of any type of retail, service, commodity, manufacturing, or processing operation, it must be optimized. As we will discuss later in this chapter, overstocking or excessive level of inventory is just as economically unfavorable as the stock shortages or insufficient inventory. Therefore, energy engineers serving the utility companies and energy consuming institutions and organizations, need to have an appreciation of the role inventory plays in the economic viability of facilities, measures and projects.

As we embark on considering some of the quantitative analytical tools and concepts associated with inventory, we need to take note of the following principles and clarifications:

- Level of inventory maintained is a function of anticipated demand.
- Higher levels of demand uncertainty result in larger volumes of (buffer) stocks that must be kept on hand.
- When required, stocks are sometimes built up to meet seasonal or cyclical demand.
- Large amounts of inventory are sometimes purchased to take advantage of discounts and economies of scale.
- Under certain circumstances, finished parts inventories are built up to meet customer demand in event of work stoppage.

CARRYING COST

Carrying cost is the cost of holding items in storage. Carrying cost can be explained further through the following points:

- Carrying costs vary with level of inventory and sometimes with length of time held.
- Carrying costs include facility operating costs (i.e. utility expenses, associated labor, etc.), overhead, record keeping, interest/finance charge, etc.
- Carrying costs are assigned on a per unit basis per time period, or as percentage of average inventory value. Carrying cost is estimated to range between 10% to 40% of the average inventory value.

SHORTAGE AND STOCK-OUT COSTS

Shortage costs, or stock-out costs, are costs associated with insufficient inventory.

- Shortage or stock-outs can result in permanent loss of sales and profits for items not on hand. Therefore, economic or financial impact of stock-outs can be quantified in terms of associated opportunity cost.

- Shortage or stock-outs can, sometimes, result in penalties. If customer is internal, work delays could be experienced resulting in opportunity cost.

INVENTORY CONTROL SYSTEMS

An inventory control system controls the level of inventory by determining how much and when to order. Software for such systems can be of-the-shelf, for smaller firms, or custom, for larger firms, organizations and institutions. Inventory control systems are premised on the following principles and concepts:

- There are two basic types of systems—periodic (fixed-time) and continuous (fixed-order quantity).

- In a periodic system, an order is placed for a variable amount after a specified period of time.

- In a continuous system, an order is placed for the same **constant amount** when inventory decreases to a specified level. This model is depicted in Figure 8-1. A continuous inventory control system, as depicted in Figure 8-1 is premised on the following assumptions:
  —Demand is known with certainty and is relatively constant over time.
  —The order quantity is received and available for use in one batch.
  —No shortage or stock-out is permitted.
  —The **lead time** for orders is constant.
  —When the inventory drops to a predetermined level, the **reorder point**, an order is placed for a fixed amount to replenish the stock.
  —The fixed amount is termed the **economic order quantity**, or EOQ. The magnitude of EOQ is set at a level that minimizes the total inventory carrying, ordering, and shortage costs.
  —The ultimate objective is to minimize incoming goods (raw materi-

als) and finished goods inventory without increasing the probability of stock-outs.

As apparent in Figure 8-1, in a continuous inventory control system, the known, constant, lead time and the inventory demand rate determine the reorder point, in terms of inventory level. The reorder point is the level of inventory at which the lead time and the demand rate intersect. Note that the units for the time axis of the inventory order cycle diagram are days, in the model shown. The time units can be weeks, months or years, just as well. Of course, in this ideal model, there is no room for error; in that, the ordered quantity arrives just in time of depletion of the last unit in inventory. Even though we are referring to the inventory in terms of units, i.e. nuts, bolts, prefabricated parts, resistors, transistors, pumps, engine assemblies, etc., this continuous inventory control system model could be just as applicable to coal shipments to a electric power generating plant. In the later case, the

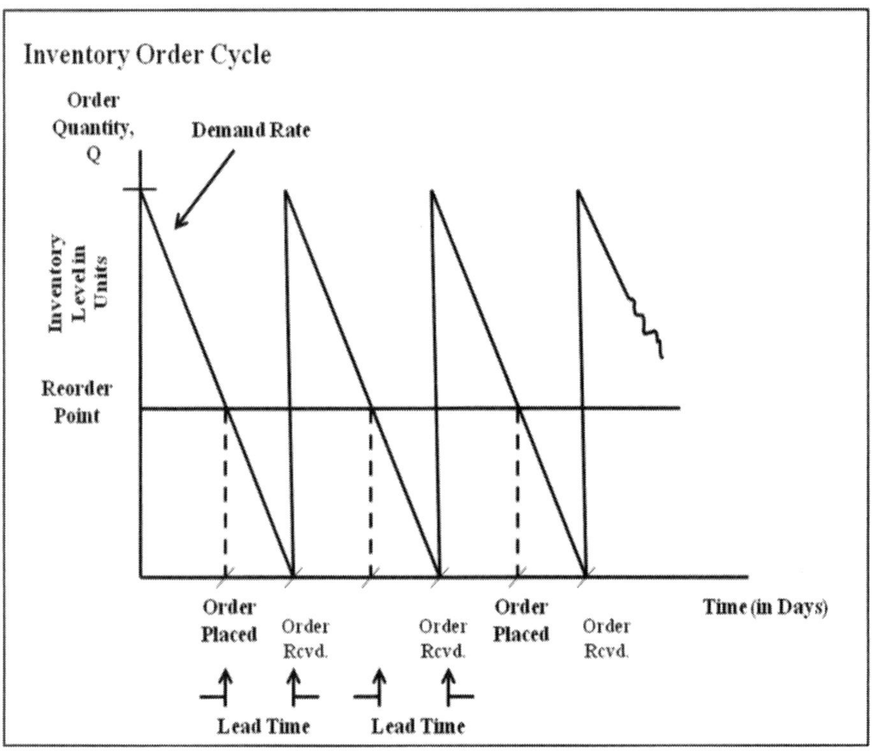

Figure 8-1. Inventory Order Cycle

reorder point would be the level of inventory of coal on premises, at which an order is issued to the coal supplier and the transport company.

EOQ, ECONOMIC ORDER QUANTITY MODEL

Figure 8-2 depicts the EOQ, economic order quantity model. This model illustrates the relationship between the carrying cost, ordering cost, order quantity, total cost and the order quantity. This model yields the following two results:

1) The optimal order quantity, or EOQ, economic order quantity.
2) The minimum total cost.

As we examine the EOQ model we need to recognize the following concepts and entities that comprise the model:

- **Annual carrying cost** is defined as the product of carrying cost per unit per year and the average inventory level:

$$\text{Annual Carrying Cost} = C_c \cdot Q/2$$

Where,

C_c = Carrying cost per unit per year, and
$Q/2$ = Average inventory

- **Total annual ordering cost** is defined as the product of cost per order and the number of orders per year.

$$\text{Annual Ordering Cost} = C_o \cdot D/Q$$

Where,

C_o = Cost per order, and
D/Q = Number of orders per year, with known and constant demand, D, and where Q is the order size

Note that Q is a variable while C_o and D are considered constant parameters.

Total annual inventory cost is sum of ordering and carrying cost:

$$TC = Co \frac{D}{Q} + Cc \frac{Q}{2} \tag{8-1}$$

As we examine the EOQ model in Figure 8-2, we notice that the EOQ occurs where the total cost curve is at a minimum value and carrying cost equals ordering cost.

$$TCmin = Co \frac{CoD}{Q_{opt}} + Cc \frac{Q_{opt}}{2} \tag{8-2}$$

$$Q\ opt = \sqrt{\frac{2\ CoD}{Cc}} \tag{8-3}$$

TC_{min} = Minimum total annual cost of inventory
Q_{opt} = Optimum order quantity or optimal order size

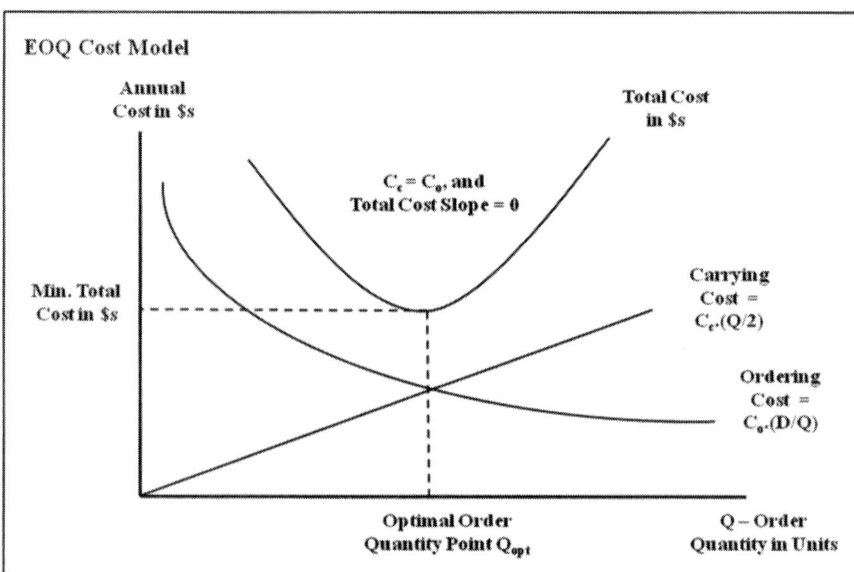

Figure 8-2. EOQ, Economic Order Quantity Cost Model

INVENTORY BASED COSTING TECHNIQUES

In production processes where raw material stock consists of a combination of recently acquired inventory as well as raw material from past shipments, firms often find themselves in a dilemma on what cost to assign to raw materials, components, supplies, etc. as they transition from **inventory into production**. When your inventory is mixed, the object is to assign cost that yields the most favorable economic results for the firm. This leads to the discussion on costing techniques.

Costing technique or method employed determines whether to use the actual purchase price, current market price or some other standard value. The costing method adopted depends on the cost flow assumptions. Two cost flow assumptions commonly employed are as follows:

- FIFO, first-in, first-out cost flow assumption
- LIFO, last-in, first-out cost flow assumption

The **FIFO** approach, assigns the oldest price first, then the next oldest and so on. During periods of rising prices, FIFO method results in **higher (inflated) net income**, relative to the LIFO method.

The **LIFO** approach, assigns the most recent price first, then the next newest and so on. During periods of rising prices, LIFO method results in **lower (suppressed) net income**, relative to the FIFO method. The LIFO method also causes the ending inventory value to be the lower than under FIFO. Due to the fact that LIFO method has the effect of suppressing net income, because of "tax" reasons, **businesses tend to apply the LIFO method during periods of rising prices.**

WORK-IN-PROCESS, WIP, INVENTORIES

Work-in-process is material to which some direct labor or overhead has been applied. Some important observations pertaining to WIP are as follows:

- Unlike raw materials and component parts, WIP have no market value.
- Like finished goods inventory, WIP is considered an asset; to be expensed only after finished goods are actually sold.

- Since WIP is considered an asset, increase in WIP tends to increase the asset base and, hence, reduces the ROI.

JUST IN TIME, JIT

JIT, just in time, is an inventory strategy that improves ROI, return on investment, of a business by minimizing finished goods inventory, raw materials inventory, WIP, and associated inventory carrying costs. In other words, high WIP would be incongruent with a JIT system. In a JIT system, the process relies on signals between different points in the process, which tell production when to make the next part. JIT also tends to improve process quality and efficiency in a manufacturing organization.

INVENTORY TURNOVER RATIO

In addition to the definition of inventory turnover ratio presented in Chapter 6, it can be viewed as a measure of the relative size of inventory as compared to the volume of cost of goods sold.

$$\text{ITR} = \frac{\text{Cost of Goods Sold (\$)}}{\text{Average Inventory (\$)}}$$

In perishable goods industry ITR tends to be high as compared to the durable goods industry ITR. For example, ITR in fruit and produce industry is once a week, 52 per year, or simply 52.

CASE STUDY 8-1—COAL FIRED ELECTRIC POWER GENERATING PLANT INVENTORY OPTIMIZATION

A coal fired electric power generating plant operates 365 days a year, on a 24/7 basis. The plant is located in West Virginia, with several coal mines and coal suppliers within a 100-mile radius. Because of a history of prices fluctuations, strikes and past supply interruptions, the firm has traditionally stock piled large inventory of coal on premises without conducting an EOQ inventory analysis. The plant produces 500 MW (Megawatt) of power. The plant uses 1.5 million tons of coal per year. The ordering cost for coal is $150 per order. The carrying cost for coal inventory on premises is $12 per ton, per year. The plant needs to determine the following in an effort to optimize the coal inventory, ordering cost and carrying cost:

a) EOQ, or optimal order size, for coal supply to the plant.
b) Total Annual Inventory Cost.
c) Number of orders to be placed annually.
d) Time between orders, or Order Cycle, in days.

Solution:

a) EOQ:

Given/Derived Data:

D = Demand = 1,500,000 tons per year
C_o = Cost per order = $150/order
C_c = Carrying Cost per ton = $12/ton

From Eq. 8-3:

$$EOQ = Q_{opt} = \sqrt{\frac{2\ CoD}{Cc}}$$

$$EOQ = Q_{opt} = \sqrt{\frac{2(\$150)\cdot(1{,}500{,}000)}{\$12}}$$

$$EOQ = Q_{opt} = 6{,}124 \text{ tons}$$

b) Total Annual Inventory Cost:

$$TC \min = \frac{C_o D}{Q_{opt}} + C_c \frac{Q_{opt}}{2}$$

$$TC \min = \frac{(\$150)\cdot(1{,}500{,}000)}{6{,}124} + \$12 \quad \frac{6{,}124}{2}$$

Total Annual Inv. Cost = $ 73,485

c) Number of orders to be placed annually = D/Q_{opt} = (1,500,000 tons per year/6,124 tons) = 245

d) Time between orders, or order cycle, in days = Q opt/D
= (6,124 tons/1,500,000 tons per year) x 365 days/year
= 1.5 days

Figure 8-3 depicts Case Study 8-1 inventory cost model in its graphical form.

SELF ASSESSMENT PROBLEMS & QUESTIONS—CHAPTER 8

1. The inventory turnover ratio based on 2004 data for ABC Corp is:

   A. 3.06
   B. 4.06
   C. 3.30
   D. None of the above

2. WIP, work In process, amplifies the asset base of business, and lowers the ROI.

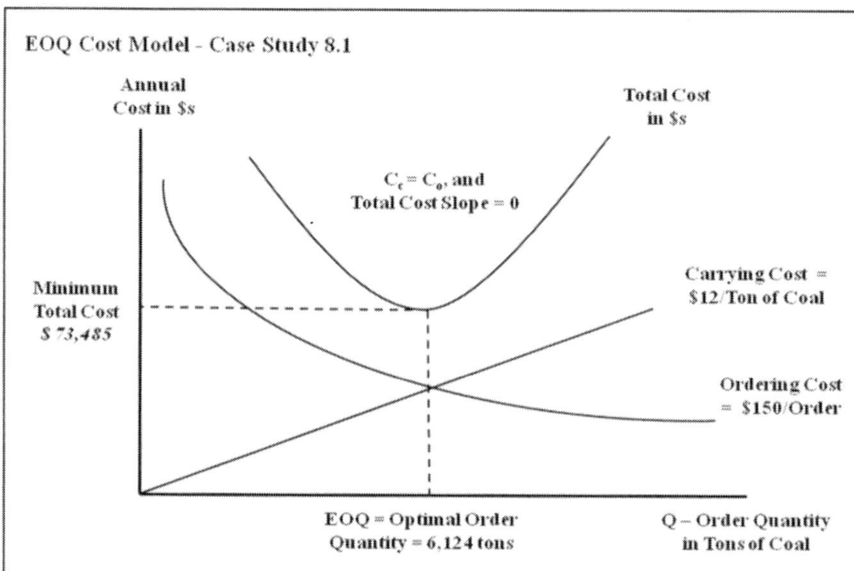

Figure 8-3. EOQ, Economic Order Quantity, Cost Model for Case Study 8-1

A. True
B. False

3. Inventory consists, mostly, of raw materials.

A. True
B. False

4. Which of the following statements is/are true about reducing spare parts inventory in a manufacturing facility?

A. It offers an opportunity to purge out obsolete spare parts.
B. It reduces the firm's inventory carrying cost.
C. It reduces the firm's asset base thus improving the ROI.
D. All of the above

5. EOQ, economic order quantity, is the order quantity at which the total inventory cost is minimum.

A. True
B. False

6. During periods of deflation or dropping prices, FIFO method for cost assignment would result in "artificially" inflated profits and higher income tax.

A. True
B. False

7. WIP, work in process, does not have definite market value and cannot be liquidated if needed.

A. True
B. False

8. A firm's inventory carrying cost is \$100/unit/year, the ordering cost is \$5.00/order/year and the demand, D, is 2,000,000 units. Calculate EOQ for this scenario.

A. 3,000 units
B. 100 units

C. 316 units
D. 447 units

9. VITA Energy sells small commercial photovoltaic panels. The cost of manufacturing these panels is $130 each and VITA sells them for $250 each. The annual sales volume is 10,000 panels. The cost of placing an order is $60, and the holding cost is $30 per unit per year. There are 360 working days per year and the lead time is five days. The EOQ for these panels is 400. This firm operates 300 days per year. Calculate the following:

(a) The total annual carrying cost for the PV panels.
A. $9,500
B. $3,000
C. $6,000
D. $1,500

(b) The annual ordering cost.
A. $2,500
B. $3,700
C. $640
D. $1,500

(c) The order cycle.
A. 12 days
B. 30 days
C. 5 days
D. None of the above

10. In Case Study 8-1, assume that one railcar or large tandem truck holds 20 tons of coal. What would the delivery be in terms of the number of trucks or railcars?

A. 20 trucks
B. 300 trucks
C. 1,000 trucks
D. 205 trucks

Chapter 9

Electric and Gas Bill Calculations and Verification

INTRODUCTION

Measurement, verification and computation of utility bills are key elements in understanding, planning for and managing the consumption of various utilities. Conservation of utilities is an essential and strategic component of overall utility cost reduction effort. Its primary impact is through enhancement of utility productivity; or, in the strict energy realm, through maximization of energy productivity. The other vital determinants of the overall cost of utilities are contract and rate structure. In a given billing month, once the utility usage has been recorded, it must be processed using the rate schedule or structure the contract calls for.

Two major forms of utility that an energy engineer must have grasp of, in residential, institutional, commercial and industrial realms, are electricity and gas. Therefore, in this chapter we will focus on mainstream approaches to computation of electricity and gas bills in the residential, commercial and industrial markets. The residential and commercial rate structures and computation methods are relatively straightforward, consequently, our discussion on residential electric and commercial gas bills will be brief. The thrust of our discussion, in this chapter, will entail analysis of electricity and gas bill computation in the industrial and commercial realm. Electric companies, gas utilities and municipal entities have various rate structures and contract alternatives for various customer categories. Some of the categories are as follows:

ELECTRIC RATE SCHEDULES

- Residential
- Commercial

- Industrial
- Uninterruptible
- Interruptible
- Hourly Pricing Rate—DSM Program Based
- OPT, Optional, Time of the Day, or, Time of Use
- Municipal or Co-op
- Onsite Generation
- Standby Generation
- Power Generator or PG schedule
- Cogeneration or PP, Purchased Power

COMMERCIAL AND INDUSTRIAL NATURAL GAS RATE SCHEDULES

- Small Commercial—Rate GS
- Large Commercial/Small Industrial—Rate CIS
- Large Volume Firm Sales Service >12,000 Mcf/Year—Rate LVS
- Interruptible Flexibility Priced Sales Service—Rate FS
- Transportation Service >12,000 Mcf/Year—Rate TS
- Very Large Volume/High Load Factor Transportation Service >5000 Mcf/Month
- Very Large Volume/High Load Factor/Sales Service >5000 Mcf/Month—Rate LVS2

As we surf the websites for various utilities, we can't help but notice the trend toward provision of "bill calculators," at these websites. While these bill calculators are quite facile and are designed, mostly for residential or small industrial consumers, they do not reveal the procedure or mechanism behind the calculations. In this chapter, we will examine the mechanics and abstracts behind some of the bill calculations. The scope of our discussion, in the case study and examples, will include a few mainstream examples representing different regions of the United States and are intended to build the reader's aptitude and skills for effective understanding and analysis of electricity and gas bills, regardless of the type of contract, rate schedule or regional differences. An energy engineer's portfolio of analytical tools would be incomplete without these skills, regardless of whether that engineer represents the energy production segment or the energy consumption arena.

RESIDENTIAL ELECTRIC BILL

As one examines the residential bills from across the various regions in the United States, one notices, that while the appearance and the format of these bills differs from region to region, the overall content and computational procedure are almost the same.

CASE STUDY 9-1

Spreadsheet 9-1 represents a residential bill in southeastern region of the US.

Spreadsheet 9-1. Residential Electric Bill

Customary/Essential Portion of Bill:	Previous	Present	Usage &Rates	Cost in Dollars
Billing Period:	2/24/10	3/22/10	28 Days	
Meter Reading, Previous	42,105	43,382	43,382 – 42,105 = 1,277 kWh	
Energy Cost Rate—Not Typically Indicated on the Bill			$0.0927 Per kWh	
RS—Residential Service—Energy Cost for the Month:				= 1,277kWh x $0.0927/kWh = $118.33
Renewable Energy Rider (Optional, for energy produced from renewable sources):				$0.16
Tax (Tax rate is typically not stated on the bill):			3%	$3.55
Total Amount Due (Bill for the month):				$118.33+$0.16 +$3.55 = $122.04

Ancillary Information Included in Some Bills for Energy Awareness Purposes

Ancillary Information	This Month	Last Year
Total kWh:	1,277	961
Billing Days:	28	28
Average kWh/Day:	46	34
Average Cost Per Day:	$4.23	$2.96

As obvious from examination of the bill calculation in Case Study 9-1, above, some of the multipliers or rates—i.e. the energy cost rate of $0.0927/kWh and the tax rate applied—critical to the calculation of the overall bill, are not explicitly stated in most bills. Also, note the recently introduced option of "Renewable Energy Rider," offered by some utilities or energy companies. Participation in this optional rider ensures that a certain percent of the energy you use will be generated through renewable fuels. The ancillary portion, included in the monthly bills by some energy firms, are a means to enhance energy consumption awareness of residential consumers.

GAS BILL—COMMERCIAL CONSUMER

Similar to a typical residential electric bill, when one examines commercial gas bills from across the various regions in the United States, one observes, that while the appearance and the format of these bills differ from region to region, the overall content and computational procedure are almost the same. Let's examine a commercial gas bill in the mid-Atlantic region of the U.S., through Case Study 9-2. This commercial entity recorded 600 Ccf of gas usage during the billing month.

CASE STUDY 9-2

The **monthly service charge** for this rate schedule is a flat amount of $10.20.

Distribution charges are charges for each Ccf (100 Cu-ft.) of gas used by the customer. For the **first 500 Ccf** used in a month, this charge is $0.433 per Ccf. For any Ccf of gas used beyond that, the charge is **$0.423 per Ccf.**

The **purchased gas cost charge** (PGC) recovers only the cost of purchased gas. It is subject to adjustments based solely on changes in the market price of gas. Currently, **the PGC is $0.735 per Ccf** of gas used within a month.

Spreadsheet 9-2. Gas Bill, Commercial Consumer

Type of Charge	Ccf	Charge Per Ccf	Cost in Dollars
Volume of Gas Usage Recorded for the Billing Month	600		
Monthly Service Charge			$10.20
Volume (Distribution) Charge for the first 500Ccf	500	$0.433	= 500 x $0.433 =$216.5
Volume (Distribution) Charge for Additional Usage Over the First 500Ccf	100	$0.423	= 100 x $0.423 =$42.3
Purchased Gas Charge	600	$0.735	= 600 x $0.735 =$441
Sub-total			$710
Tax (Tax rate is typically not stated on the bill):		3%	= 0.03 x $710 = $21.30
Total Amount Due (Bill for the month):			$731.30

Even though, discussion of a large volume gas contract is outside of our scope, for reference, the Gas Rate Schedule for Large Volume Gas Sales Service, published by Richmond, VA, Department of Public Utilities, is appended below. Examination of this rate schedule reveals the fact that, with large gas consumers whose consumption exceeds 12,000 Mcf, determination of a monthly bill tends to become a bit more complex. Large gas contracts or schedules, typically, include the following billing components:

1. Customer Charge
2. Demand Charge
3. Transportation Charge
4. Utility Tax

The transportation charge in large volume gas schedules is, typically, tiered. The demand rate, on the other hand, is flat but determined by three mutually exclusive means, as shown in Richmond Gas Rate Schedule below.

GAS RATE SCHEDULE
AMENDED JULY 1, 2009
SCHEDULE LVS
Large Volume Gas Sales Service
(An Excerpt)

A. APPLICATION:

Service is available throughout the service territory served by the City to all firm, non-residential gas sales customers that take in excess of 12,000 Mcf of gas over a consecutive twelve month period

B. MONTHLY RATES AND CHARGES:

- Customer Charge $520.00 per month
- Demand Charge $10.62 per Mcf of Billing Demand

Transportation Charge

- First 1,500 Mcf $1.44 per Mcf
- Next 1,501 to 11,500 Mcf $0.72 per Mcf
- For all additional cubic feet over 11,500 Mcf $0.51 per Mcf

Purchased Gas Cost Charge—Weighted Average Commodity Cost of Gas ("WACCOG"). The purchased gas cost charge shall be determined each month and shall include all variable costs associated with gas manufactured by the City and all commodity charges, surcharges, tracking adjustments, and all other non-fixed charges of pipelines and gas suppliers incurred by the City. This purchased gas cost charge can include gas bought by the City at a fixed cost to serve a customer or group of customers as approved by the Director. Any agreement to fix such costs shall be specified in the service agreement.

C. DETERMINATION OF DEMAND:

The demand may, at the option of the Director of Public Utilities of the City, be determined either by measurement, by estimate or by agreement.

A. **By Measurement**—The demand in any month shall be the highest use of gas in Mcf in any period of twenty-four (24) consecutive hours as measured by the demand meter.
B. **By Estimate**—The demand in any month shall be taken as 1/20 of the Mcfs used in such month.
C. **By Agreement**—At a level to recover the upstream demand charges used to serve the customer. Such level shall be specified in the service agreement. Customer usage above this firm daily demand level shall be regarded as interruptible and will

be subject to the terms in Section 6 of the City Rate Schedule FS (City Code, Sec. 29-244. Flexibly priced interruptible gas sales service).

D. **BILLING DEMAND:**

The billing demand in any month shall be the higher of:

1. The demand as determined in such month under section C above or
2. The highest billing demand in any of the preceding months of November through April, provided, however, that for new customers or customers transferring from another rate schedule, the highest billing demand may, at the option of the Director of Public Utilities of the City, be estimated based on the proposed use of service under this rate schedule.

E. **UTILITY TAX:**

Bills rendered under this schedule shall be subject to any applicable utility tax.

LARGE INDUSTRIAL ELECTRIC RATE SCHEDULE

As we learned earlier in this chapter, residential electricity bill calculation tends to be perspicuous, involves a few simple steps and one dominant variable; which is the energy usage recorded in kWh. On the contrary, we are about to see that large industrial electrical contracts and schedules are more complex. They involve multiple calculation steps involving multiple substantial variables. These major variables are energy (on-peak and off-peak), demand—and, where applicable—HP, or hourly price component. We will analyze the calculation and composition of a large industrial bill through Case Study 9-3, involving the application of OPT, optional, time-of-use rate schedule. This case study is premised on Duke Energy's OPT, time-of-use rate schedule, appended below. For reference and comparison, an excerpt of an equivalent rate schedule offered by PNG, also known as Progress Energy, is included in this section.

CASE STUDY 9-3. ELECTRICAL POWER BILL—LARGE INDUSTRIAL POWER CONSUMER ON DUKE ENERGY GRID

Background

This case study is modeled on a scenario with large industrial customer on Duke Energy's grid, in the southeastern United States, circa 2000. The rates in \$/kW or \$/kWh and the fixed charges are assumed to represent that time frame, during non-summer months. These rates are somewhat comparable to the 2009 Duke non-summer rates shown later in this chapter under Exhibit 9-1. This consumer is assumed to have a 45 MW (megawatt) contract with Duke Energy. This facility is a 24/7,365-day operation and a significant portion of its load can be shifted from on-peak periods to off-peak periods, during its daily operations. This overall service contract of 45 MW is split into the following two components:

A. A 30 MW, uninterruptible, power service, on OPT, time-of-use schedule.

B. A 15 MW, HP, hourly pricing schedule. This schedule applies to incremental, interruptible, load and was offered to this customer as a part of DSM, demand-side management, program.

Spreadsheet 9-3 captures all of the measured data, derived billing parameters, standard charges, tiered demand charge rates, energy charge rates, and computed line item charges for the billing month. This spreadsheet shows the arithmetic behind the derived or extended line item charges.

Baseline Billing Determinants

The on peak billing demand and the billing demand are typically the same. This segment of the bill represents derived and measured data. This measured portion is in form of the energy consumption (kWh) measured in 15- or 30-minute intervals; which are then divided by 0.25 hour or 0.5 hour, respectively. This yields demand (kW) for each of the intervals for the entire billing month. The utility company then selects the highest of these demands as the peak demand or billing demand for the month. The other billing determinants consist of the total energy usage during on-peak and off periods. The consumed energy data is in measured and recorded for the billing month, in kWh.

Basic Facilities Charge

This charge could be considered to represent administrative cost associated with the generation and processing of the bill. This charge stays, relatively, constant over time.

Extra Facilities Charge

This charge is a means for the utility to recoup its cost embedded in providing "extra facilities" to the customer. In this case study, the extra facili-

Spreadsheet 9-3. Large Industrial Electric Bill OPT Schedule

Description	Demand and Energy Parameters	Rates	
Baseline Billing Determinants			
Billing Demand			26,000 kW
On-Peak Billing Demand			26,000 kW
On-Peak Energy Usage			4,334,573 kWh
Off-Peak Energy Usage			14,500,000 kWh
Baseline Charges			
Basic Facilities Charge			$36.07
Extra Facilities Charge			$13,000
On-Peak Billing Demand Charge			
For the First 2,000 kW	2,000 x	$7.64 per kW =	$15,280.00
For the Next 3,000 kW	3,000 x	$6.54 per kW =	$19,620.00
For All Over 5,000 kW	21,000 x	$5.43 per kW =	$114,030
Economy Demand Charge	0 x	$1.03 per kW =	$0
On-Peak Energy Usage	4,334,573 x	$0.042393 Per kWh =	$183,755.55
Off-Peak Energy Usage	14,500,000 x	$0.021039 Per kWh =	$305,065.50
Total Baseline Charge		$0.033135 Per kWh	**$650,787.12**
HP Billing Determinants			
Actual Demand	40,000 kW		
HP Charges			
Incremental Demand Charge	14,000 kW x	$0.26 Per kW=	$3,640.00
New Load x Avg. Hourly Price	8,805,800 x	$0.021672 Per kWh =	$190,839.30
			$194,479.3
Total Bill for the Month			**$845,266.42**

ties consisted of a set of redundant transmission lines installed by Duke to enhance reliability of the power service to this customer. In some cases, this type of charge is associated with the upgrade of main switch yard step down transformers, regulators, separate metering, etc.

On-peak Billing Demand Charge

The overall billing demand of 26,000 kW is tiered into three segments: first 2,000 kW, next 3,000 kW and the remaining 21,000 kW. Each of these demand tiers are applied respective charges, in $/kW, as shown in Spreadsheet 9-3, to yield tiered line item demand charges. These tiered line item demand charges are later added to other bill line item charges.

Economy Demand Charge

This charge is triggered when the total demand of the facility falls more than 50% below the demand (kW or MW) specified in the current contract. When economy demand is triggered, the utility takes the difference between the demand equivalent of 50% of the contracted demand and the actual demand (below the 50% level) and multiplies it with the economy demand rate. In this case study, the economy demand charge rate is assumed to be $1.03 per kW. Since in this case study the demand is well above 50% of the contract level, economy demand is not triggered and, therefore, the economy demand charge is zero.

Energy Charge

This charge accounts for the actual, measured, energy consumed during the billing month. This charge comprises of two components as shown in Spreadsheet 9-3; the on-peak energy charge and the off-peak energy charge. As apparent from examination of the spreadsheet, the on-peak rate ($0.042/kWh) is almost twice the magnitude of the off-peak rate ($0.021/kWh).

Total Baseline Charge

The total baseline charge is simply the sum of all line items calculated or identified to that point; $650,787.12, in this case.

HP, Hourly Pricing, Billing Determinants

This portion of the bill represents the demand and energy charges associated with the load that fall under the Hourly Pricing contract. The hourly pricing demand was measured and billed at 14,000 kW, in this case. The energy consumed under the hourly pricing contract was 8,805,800 kWh.

HP Charges

Special HP rates are applied to the recorded demand and energy under HP contract, yielding respective line item charges as shown on Spreadsheet 9-3.

Total Bill for the Month

The last line of the bill represents the sum of total baseline charge and the HP charge, amounting to the total bill of **$845,266.42** for the month.

Exhibit 9-1

Duke Energy Carolinas, LLC

Electricity No. 4
North Carolina Sixth Revised Leaf No. 23
Superseding North Carolina Fifth Revised Leaf No. 23

SCHEDULE OPT-G (NC)
OPTIONAL POWER SERVICE, TIME OF USE
GENERAL SERVICE

Effective for service on and after January 1, 2010 NCUC Docket No. E-7, Sub 909, Orders dated December 7, 2009 and December 28, 2009 Page 1 of 3

AVAILABILITY (North Carolina Only)
Available to the individual customer.

Service under this Schedule shall be used solely by the contracting Customer in a single enterprise, located entirely on a single, contiguous premises.

This Schedule is not available to the individual customer who qualifies for a residential schedule or industrial schedule, nor for auxiliary or breakdown service. Power delivered under this schedule shall not be used for resale or exchange or in parallel with other electric power or as a substitute for power contracted for or which may be contracted for, under any other schedule of the Company, except at the option of the Company, or for service in conjunction with Rider SCG or Rider NM, under special terms and conditions expressed in writing in the contract with the customer.

The obligations of the Company in regard to supplying power are dependent upon its securing and retaining all necessary rights-of-way, privileges, franchises and permits, for the delivery of such power. The Company shall not be liable to any customer or applicant for power in the event it is delayed in or is prevented from, furnishing the power by its failure to secure and retain such rights-of-way, rights, privileges, franchises and permits.

TYPE OF SERVICE

The Company will furnish 60 Hertz service through one meter, at one delivery point, at one of the following approximate voltages, where available:

Single-phase, 120/240 volts, 120/208 volts, 240/480 volts or other available single-phase voltages at the company's option; or

- 3-phase, 208Y/120 volts, 460Y/265 volts, 480Y/277 volts; or
- 3-phase, 3-wire, 240, 460, 480, 575, or 2300 volts; or
- 3-phase, 4160Y/2400, 12470Y/7200, or 24940Y/14400 volts; or
- 3-phase voltages other than those listed above may be available at the Company's option if the size of the Customer's contract warrants a substation solely to serve that Customer, and if the Customer furnishes suitable outdoor space on the premises to accommodate a ground-type transformer installation, or substation, or a transformer vault built in accordance with the Company's specifications.

The type of service supplied will depend upon the voltage available. Prospective customers should determine the available voltage by contacting the nearest office of the Company before purchasing equipment.

Motors of less than 5 H.P. may be single-phase. All motors of more than 5 H.P. must be equipped with starting compensators. The Company reserves the right, when in its opinion the installation would not be detrimental to the service of the Company, to permit other types of motors.

RATE: I.	**Basic Facilities Charge per month**	**$36.51**	
II.	**Demand Charge**	**Summer Months**	**Winter Months**
A. On-Peak Demand Charge		June 1-September 30	October 1-May 31
For the first 2000 kW of Billing Demand per month, per kW		$14.0080	$8.2460
For the next 3000 kW of Billing Demand per month, per kW		$12.8319	$7.0591
For all over 5000 kW of Billing Demand per month, per kW		$11.6449	$5.8613
B. Economy Demand Charge		$1.1136	$1.1136

III. Energy Charges	**All Months**
A. All On-Peak Energy Per Month, Per kWh	5.7026 ¢
B. All Off-Peak Energy, Per Month, Per kWh	3.3994 ¢

Exhibit 9-2

Carolina Power & Light Company 17
d/b/a Progress Energy Carolinas, Inc.
(North Carolina Only)
(Excerpt)

LARGE GENERAL SERVICE
(TIME-OF-USE)
SCHEDULE LGS-TOU-15
AVAILABILITY

This Schedule is available on a voluntary basis for electric service used by a nonresidential customer with either a Contract Demand that equals or exceeds 1,000 kW or whenever the registered or computed demand equals or exceeds 1,000 kW in the preceding 12 months.

This Schedule is not available: (1) for breakdown, standby, or supplementary service, unless used in conjunction with the applicable standby or generation service rider for a continuous period of not less than one year; (2) for resale service; (3) for short-term or temporary service; or (4) for any new customer with a Contract Demand in excess of 100,000 kW.

APPLICABILITY
This Schedule is applicable to all electric service of the same available type supplied to Customer's premises at one point of delivery through one meter.

TYPE OF SERVICE

The types of service to which this Schedule is applicable are alternating current, 60 hertz, three-phase 3 or 4 wires, at Company's standard voltages of 480 volts or higher or the voltage at which Customer was being served on September 19, 1983. When Customer desires two or more types of service, which types can be supplied from a three-phase 4 wire type, without voltage transformation, only the one of these two types necessary for Customer's requirements will be supplied.

CONTRACT DEMAND

The Contract Demand shall be the kW of demand specified in the Service Agreement.

MONTHLY RATE

I. Basic Customer Charge:
$500.00

II. kW Demand Charge:

Service Rendered During the Calendar Months Of:
June through September October through May

A. On-Peak Billing Demand:

First 5,000 kW of
Billing Demand $19.56 per kW $14.25 per kW

For the next 5,000 kW
of Billing Demand $18.56 per kW $13.25 per kW

All over 10,000 kW of
Billing Demand $17.56 per kW $12.25 per kW

B. All off-peak excess
Billing Demand $ 1.00 per kW $ 1.00 per kW LGS-TOU-15 Sheet 2 of 4

III. kWh Energy Charge:

4.879¢ per on-peak kWh
4.379¢ per off-peak kWh

IV. Renewable Energy Portfolio Standard (REPS) Adjustment:
The monthly bill shall include a REPS Adjustment based upon the revenue classification:
Commercial/Governmental Classification – $3.22/month
Industrial/Public Authority Classification – $32.20/month

Upon written request, only one REPS Adjustment shall apply to each premise serving the same customer for all accounts of the same revenue classification. If a customer has accounts which serve in an auxiliary role to a main account on the same premise, no REPS charge should apply to the auxiliary accounts regardless of their revenue classification (see Annual Billing Adjustments Rider BA).

V. Transformation Discounts:
When Customer owns the step-down transformation and all other facilities beyond the transformation which Company would normally own, except Company's metering equipment, the charge per kW of on-peak Billing Demand and per kWh will be reduced in accordance with the following:

Transmission Service Distribution Service

Transformation Discount Transformation Discount
$0.48/kW $0.75/kW
$0.0008/kWh $0.0001/kWh

Transmission: For Customer to qualify for the Transmission Service Transformation Discount, Customer must own the step-down transformation and all other facilities beyond the transformation which Company would normally own, except Company's metering equipment, necessary to take service at the voltage of the 69 kV, 115 kV, or 230 kV transmission line from which Customer received service.

SELF ASSESSMENT PROBLEMS AND QUESTIONS—CHAPTER 9

1. HP, or hourly pricing, program is a standard feature of all OPT, or time-of-use, contracts or schedules.

   A. True
   B. False

2. The tiered demand charge rate structure with electrical OPT schedule is similar to the tiered transportation cost rate structure of the large natural gas schedules.

   A. True
   B. False

3. Transportation cost is one of the key components of residential gas bills.

   A. True
   B. False

4. A large industrial electricity consumer set the peak demand in a given billing month at 40 megawatt. The demand rate structure is same as that included in PNG time-of-use rate schedule, as shown in Exhibit 9-2. The demand cost for the month would be:

   A. $367,000
   B. $505,000
   C. $407,000
   D. None of the above

5. If the Duke Energy OPT, Time of Use, 2009 Schedule, as shown in Exhibit 9-1 had been used in Case Study 9-3, for the month of July, with

no HP energy usage during the month, and an extra facilities charge of $13,000, the total bill for the month would be:

A. $967,000
B. $585,000
C. $1,064,187
D. None of the above

6. If the gas usage in Case Study 9-2 were reduced to 550 Ccf, the total gas bill for the month would most nearly be:

A. $672,000
B. $525,000
C. $822,000
D. None of the above

Chapter 10

Types of Cost, Life Cycle Cost and Repair vs. Replace Decisions

INTRODUCTION

As we have learned through various topics in this text, profits and, ultimately, the viability, growth and survival of firms, organizations and institutions depend innately on cost of producing goods and services. Often, when price pressures limit and suppress revenues, business entities turn to costs in order to survive and be viable. In recessionary periods and in the face of stern competition, cost reduction efforts become paramount. However, as with many other offsetting variables in business environment, one cannot minimize cost without fully understanding of its characteristics, forms, and its relationship to other variables that determine the bottom line. The realm of energy is no different than any other business when it comes to the cost, profitability and competitive advantage. Whether we are contending with initial costs or energy projects or the cost of operating existing energy assets, we can only control cost effectively if we understand it and the roles it plays, in its various forms, in offsetting the revenues in our quest for profits. Some of the cost terms we will discuss may not have a direct relationship to energy. But, since energy touches every aspect of business and commerce, as an energy engineer you would need to understand the cost structure and composition of your clients, internal and external.

Therefore, in this chapter we will explore different types of costs. Most cost categories we will consider in this chapter are profound and could require prolonged discussion and analysis for a thorough examination. However, in this text we will discuss most of the cost concepts briefly and then, through use of short examples, we will illustrate those concepts. We will conclude this chapter with a case study on the topic of life cycle cost analysis.

Cost Flow

In a manufacturing environment, all manufacturing costs—material, labor and overhead –become assets when charged to production.

First these costs constitute the cost of the work-in-process inventory.

Later these costs constitute the cost of the finished goods inventory.

Manufacturing costs do not become expenses until the time the finished goods are sold. In the power generation realm, the cost of flows through the operations in form of the cost of fuels, labor and services; ultimately, resulting in electrical energy as product. Since the price of electrical energy cannot, in most cases, be changed instantaneously, any fluctuation in cost must be absorbed by equivalent recession of net income.

Cost Variances

Cost variances consist of costs that fall above or below budget.

A negative variance consists of actual cost in excess of the budget.

A positive variance consists of the difference between actual costs and the budget when the actual costs are less than the budget.

A positive variance is also referred to as a favorable variance. As an energy engineer, with stewardship an accountability for energy cost in a plant, one has to explain significant variances against the budget, regardless of whether they are positive or negative. A successful energy or utilities engineer is one who can utilize historical data load profiles and production projections to predict cost accurately and thus formulate budgets that are more on the mark. Well formulated budgets limit operational cost variances, in the absence of uncontrollable, force majeure events.

Direct Labor Costs

Direct labor costs are costs associated with manufacture, fabrication or assembly of product. Direct labor costs exclude times that laborers are idle. Direct labor costs at a natural gas distribution plant would be limited to the wages of the plant operators, maintenance technicians and other employees who "touch," control or add value to the product.

Direct Material Costs

Direct material costs are costs associated with raw materials or components required for the manufacture, fabrication or assembly of product. Direct material costs in a coal fired power generating plant would include the cost of coal (delivered), utilities such as water, consumable effluent filtration slurry ingredients and the cost any other consumables that are required in the combustion of coal and conversion of steam energy into electrical energy.

Discretionary Cost

Discretionary cost is the cost that is subject to control by management. For instance, the amount appropriated for advertising budget. The funds spent by an energy company in a public opinion restoration campaign, after a substantial environmental discharge event, would constitute discretionary cost.

Fixed Costs

Fixed costs are costs that remain constant regardless of number of units produced. If a renewable energy windmill farm is built upon federal land, through a lease, the lease payments, annual or monthly, would constitute fixed costs.

Indirect Cost

Indirect cost is a cost incurred to serve more than one segment of an organization. For instance, depreciation expense on a building that houses a bank of micro-turbines and some production equipment, would be an indirect cost on the books of the manufacturing entity.

Incremental Cost

Indirect cost is the added cost of producing one more unit of product. If power company determines that it would need to expend $0.01 to produce an additional kWh, the incremental cost would be $0.01/kWh.

Overhead Costs

Overhead costs consist of all indirect and fixed costs. Wages for staff functions such as front line management, administrative associates, sales functions, accounting functions, costs of employee benefits, etc. are examples of overhead costs. In a large CSP, concentrated solar power, facility, overhead costs would include: salaries of shift operation supervision, plant engineer, operations manager, maintenance manager, safety manager and administrative associates. Any benefits or perks provided to the employees, such as subsidized meals, due to the remoteness of the facility, would be considered as overhead cost.

Overhead (Cost) Rate

Overhead rate is used to apply overhead cost to work in process. Overhead rate is determined by dividing the estimated overhead cost by the estimated activity level (in dollars or other units). If a nuclear power plant has annual overhead cost of $100,000, and it generates 500 megawatts of power, one measure for overhead cost rate in that case would be $200/MW/year.

Sunk Cost

A sunk cost is cost that has already been incurred, is irreversible and unchangeable. For instance, an expenditure of $100,000 for the acquisition of a direct fuel cell power generating system would be categorized as sunk cost; unless, this system is acquired through a short term lease, on trial basis.

Uncontrollable Cost

An uncontrollable cost is a cost that cannot be controlled by a manager who incurs or is responsible for that cost. PPV, or purchase price variance, associated with the natural gas and electricity procurement is a example of incontrollable cost in the energy area. When the actual cost of fuels or electricity exceeds the cost rate projected or budgeted for the fiscal year—or for months within a fiscal year if the rates are seasonal, i.e. the time-of-use, OPT, electricity cost rates—the difference between the actual rates and the budgeted rates is referred to as PPV, purchased price variance.

Variable Cost

Variable cost is the cost per unit or cost associated with the production of one unit of product. Let's assume that Richmond pays a total price of $0.30 per Ccf for natural gas that it distributes and sells to its commercial consumers, in Chapter 9, Case Study 9-2. The variable cost in that case would be $0.30 per Ccf.

COST EXAMPLES

Now that we have a basic understanding of some of the cost concepts, lets reinforce hone our comprehension through some examples that extend beyond the energy realm.

Examples 10-1. Direct/Indirect Material Cost

When a craftsman in a furniture manufacturing plant orders 50 tool sharpening stones each year, it would constitute indirect material cost because the stones are for the purpose of maintaining tools used to perform the work. Conversely, if the materials in question were wood, cloth, nails, staples, glue, fabric, etc., they would represent direct material costs.

Examples 10-2. Direct/Indirect Material Labor

A software developing firm employs 30 programmers, 3 project managers, a human resource manager, a president/CEO and one administrative associate. The human resource manager's salary, in this business scenario, would constitute indirect labor. A programmer's work, on the other hand, would be considered direct labor because his/her effort contributes directly toward the creation of the product.

Examples 10-3. Discretionary Cost

The president in the software developing firm referenced above decides to hold a golf outing for his staff. The cost for this event would fall under the discretionary cost category.

Examples 10-4. Fixed and Incremental Costs

The software firm referenced above measures its output in terms of projects completed per month. Due to an increase in demand and expansion of clientele, this firm hires another programmer at a salary of $72,000 per year. By virtue of this addition to the staff, the firm is expected to increase its output by one project per month. In this case the firm's per unit incremental cost is $6,000 and the firm's fixed cost would rise by $6,000/month. By definition, the incremental cost, or the cost for completing an additional project is $72,000/12 = $6,000/month. Since, in this case, the firm is obligated to pay an additional $6,000 per month to the new programmer, continuously, for an indefinite period of time, the firm's fixed cost or obligation would rise by $6,000.

Examples 10-5. Sunk Costs

A simple example of sunk cost could be a new industrial grade copier that software firm, referenced above, purchases, provided of course, that it is not leased.

Examples 10-6. Negative Variance

Consider the furniture company from Example 10-1.The production manager in the assembly department exceeds the budget for the month of March by $11,000. The negative variance in this case would be $11,000.

LIFE CYCLE COST AND REPAIR VS. REPLACE DECISIONS

Repair vs. replace decisions involve comparison of the life cycle cost of equipment under two mutually exclusive scenarios:

a) Life cycle cost if the equipment is replaced.
b) Life cycle cost if the equipment in question is repaired.

Life Cycle Cost

Life cycle cost of a piece of equipment or project, energy related or not, is the total cost of owning and operating that piece of equipment over its expected life. This cost, in most cases, would consist of the following:

a) Initial cost or initial investment. This could be the total turnkey cost of a project constituting an energy measure. This cost should include combined cost of equipment, installation material, installation labor, start-up and commissioning cost, etc.

b) Total preventive, predictive and other typical maintenance costs over the life span of the equipment or system.

c) Fuel, electricity or other energy cost over the life span of the equipment. This would need to take into account the efficiency of equipment.

d) Opportunity cost or lost production if the equipment malfunctions. One would need to estimate and account for opportunity cost under alternative scenarios.

e) Depreciation: D (S/L) = (Original Cost – Salvage Value)/Life in # of Years

CASE STUDY 10-1. MAKING A DECISION BETWEEN TWO (2) ALTERNATIVES, BASED ON LIFE-CYCLE COST, WITHOUT CONSIDERATION OF TIME VALUE OF MONEY

A 100 HP, 480 VAC, 3-phase blower motor has experienced a ground fault due to winding failure. This motor has been repaired or rewound twice before. The motor has lost 5% of its efficiency during each of the past two (2) rewinds, due to core losses. The maximum actual load on the motor is 90 HP. Prior maintenance history, equipment specifications, purchasing quotes, electricity cost rates and other pertinent basic assumptions are as follows:

- A typical rewind or repair service, for a 100 HP motor is $1,600.
- A new premium efficiency motor cost, approximately, $3,400.
- Electricity cost at this facility averages $0.05/kWh.
- Assume initial efficiency of 96% and a power factor of 0.9.
- Assume that the motor experiences a fault, due to the ambient conditions and duty cycle, at the 5th year, 8th year and the 10th year.
- Assume 24/7 operation and an average efficiency of 95% over 10 years. Assume that the electricity cost stays constant over 10 years.
- Assume combined Lost Production Cost (opportunity cost) and Maintenance Labor Cost of $3,000 per failure.

What is the life cycle cost, or cost of owning, maintaining and operating this motor, over a period of 10 years under the following two scenarios?

A. Scenario (1)—Rewind twice, then replace at the 3rd failure.
B. Scenario (2)—Replace upon 1st Failure.

Solution:

Energy Cost Analysis:

Original Efficiency of the Motor = **96%**
Motor Efficiency after 1st Rewind = 96% x (1– .05) = **91.2%**
Motor Efficiency after 2nd Rewind = 91.2% x (1– .05) = **86.64%**
Motor Efficiency after 3rd Rewind = 86.64 HP x (1– .05) = **82.31%**

Energy Cost, 1st 5 Years
= 100 HP x 0.746 kW/HP x 24 x 365 x 5/0.96 x 0.05 = $170,181
Annual Energy Cost = $170,181/5 = $34,036

Energy Cost, After 1st Rewind, for 3 Years
= 100 HP x 0.746 kW/HP x 24 x 365 x 3/0.912 x 0.05 = $107,483
Annual Energy Cost = $107,483/3 = $35,828

Energy Cost, After 2nd Rewind, for 2 Years
= 100 HP x 0.746 kW/HP x 24 x 365 x 2/0.8664 x 0.05 = $75,427

Scenario (1)—Rewind twice, then replace at the 3rd failure:
Life Cycle Cost = ($170,181+$107,483+$75,427) +($1,600+$1,600+$3,400)+(3)x($3,000) = $368,691

Scenario (2)—Replace upon 1st Failure:
Life Cycle Cost = ($34,036 x 10) + ($3,400+$3,400) + (2)x(3000) = **$353,163**

Conclusion: Scenario (2) or replacement of the motor at the first failure carries a lower life cycle cost, a favorable cost difference of $15,528.

CASE STUDY 10-2. MAKING A DECISION BETWEEN TWO (2) ALTERNATIVES, BASED ON LIFE-CYCLE COST, WITH CONSIDERATION OF TIME VALUE OF MONEY

Computation associated with this case study, as many preceding case studies in this text, is based on Microsoft Excel ®. Even though the Excel spreadsheet spans over the 10-year life of the motor, only years 1, 6, 8 and 10 are shown in Spreadsheet 10-2, below. These are the years when significant transactions or cash flows occur. A complete 10- year version of this spreadsheet is included in Appendix B.

As we examine Spreadsheet 10-2, we observe that due to the energy efficiency changes associated with rewinding of the motor, the energy consumption related cash flows change through the 10-year span under the rewind alternative. Conversely, the energy cost related annual cash flows, under the motor replacement scenario, stay constant at ($34,036).

Spreadsheet 10-2. Making a Decision Between Two (2) Alternatives, Based on Life-Cycle Cost, With Consideration of Time Value of Money

Case Study 10-2—Life Cycle Cost Comparison Based on Time Value of Money:

Discount Rate: 10%

Scenario (1): Rewind Twice, then Replace at 3rd Failure:

Year	1	6	8	10
Actual Energy Cost:	$(34,036)	$(35,828)	$(35,828)	$(37,713)
Purchase & Rewind Costs:	$(3,400)	$(1,600)	$(1,600)	
Lost Prod & Maint. Cost:		$(3,000)	$(3,000)	$(3,000)
Cash Flows:	$(37,436)	$(40,428)	$(40,428)	$(40,713)

NPV of All Costs: **$(223,871)**

Scenario (2): Replace Upon 1st Failure:

Year	1	6	8	10
Actual Energy Cost:	$(34,036)	$(34,036)	$(34,036)	$(34,036)
Purchase & Rewind Costs:	$(3,400)	$(3,400)		
Lost Prod & Maint. Cost:		$(3,000)		$(3,000)
Cash Flows:	$(37,436)	$(40,436)	$(34,036)	$(37,036)

NPV of All Costs: **$(216,997)**

Cost Difference, Scenario (1) vs. Scenario (2): **$6,874**

The purchase and rewind cost transactions are stated under their respective years of occurrence; ($3,400) for replacement of the motor and $1,600 for rewinding. The ($3,000) cash flow listed in the lost production or maintenance cost category appears in conjunction with each motor rewinding and replacement event, in each scenario.

After identification of all costs as negative cash flows, for all events in each cost category, they are added, for each year, to yield total annual cash flows. Each annual cash flow is then converted into its corresponding present value; this step is not shown in the spreadsheet. The sum of all present values, from each year, represents the NPV, net present value for the alternative.

The NPV values for both alternatives analyzed above are negative because they represent net costs; earnings are not pertinent in this case. As covered in the section on NPV analysis, in this text, when analyzing alternatives on the basis of cost, we must choose the alternative that yields the least negative NPV. Since the "replace" alternative offers a smaller negative NPV of ($216,597), versus a larger negative NPV of ($223,871) associated with the "rewind" option, based on this life-cycle cost analysis, it would be more economical to replace the motor instead of rewinding it.

SELF ASSESSMENT PROBLEMS & QUESTIONS—CHAPTER 10

1. In a uninterruptible power supply manufacturing facility, a custom circuit board assembly technician uses two rolls of solder, per shift, to connect electronic devices to the circuit board. The cost of these two rolls of solder constitutes indirect material cost.

   A. True
   B. False

2. A small hydroelectric power plant employs 30 machine operators, three (3) managers, a human resource manager, a president/CEO and a receptionist. The president's salary, in this business scenario, would constitute overhead.

   A. True
   B. False

3. A local cogeneration facility consumes $20,000 worth of natural gas each week. The cost of this gas would represent:

   A. Direct Cost
   B. Incremental Cost
   C. Fixed Cost

4. A coal fired power plant is fined $50,000, by the EPA, for an excessive accidental release of sulphur dioxide into the atmosphere. This cost would be categorized as:

   A. Discretionary Cost.

B. Cost Flow
C. Uncontrollable cost.

5. During a monthly cost review meeting, one of the department managers reports that the total monthly expenses in his department were more than the budgeted allowance for the month. The difference between the budget and the actual expenses recorded would constitute a:

   A. Negative Variance
   B. Positive Variance
   C. Variable Cost

6. In a gas turbine manufacturing facility the cost is tracked through 30 fabrication and assembly stations, beginning with the raw material or component warehouse, all the way to the shipping warehouse. The costing in this situation would represent:

   A. Work In Process
   B. Cost Flow
   C. Direct Cost
   D. Both A and B

7. Life cycle cost of capital investments does not include energy cost associated with the operation of equipment.

   A. True
   B. False

8. In Case Study 10-2, the NPV analysis of both alternatives, replace and rewind, yielded negative numbers due to the fact that the focus of both analysis was on energy savings.

   A. True
   B. False

9. In Case Study 10-2, the maintenance costs, including activities such as predictive maintenance and preventive maintenance, were only included in the "rewind" scenario because a new motor would not require such measures.

   A. True
   B. False

10. In Case Study 10-2, which of the following contributed most substantially in determining the annual cash flows, after the first year, for NPV formulation?

    A. Maintenance cost
    B. Initial cost
    C. Rewinding Cost
    D. Energy Cost

11. When formulating a financial decision between two or more "cost based" scenarios, one must choose the scenario pertaining to most negative NPV.

    A. True
    B. False

12. In Case Study 10-2, examine the "lost production and maintenance cost" rows, for the replace and rewind scenarios. Which scenario is likely to have smaller opportunity cost relative to unplanned downtimes?

    A. Replace Scenario $6,000
    B. Rewind Scenario $9,000

13. The transactions or costs listed under the eighth year, for the replace scenario, don't include any replace cost because the motor life expectancy is 5 years.

    A. True
    B. False

Chapter 11

Energy Performance Contracting and ESCOs—

Business, Economic and Financial Perspective; Lease and Capital Investment Alternatives

INTRODUCTION AND BRIEF HISTORY OF EPC AND ESCOs

EPC, energy performance contracting, is a service vehicle for provision of energy conservation or energy productivity enhancement services. In most cases, EPC type contracts and projects involve turnkey service. Turnkey service is defined as a comprehensive service provided by a vendor or contractor that begins with definition and scope of an energy project and ends with project start up, commissioning and subsequent verification of energy savings. Projects covered by EPC range from simple energy conservation efforts such as replacement of inefficient lighting systems to more complex projects that address the supply side of the energy equation through renewable energy systems.

EPC projects and contracts often include guarantees that the savings produced by a project will be sufficient to finance the full cost of the project. The guarantee to fund the project through the savings generated by the project is what distinguishes EPC projects from non-EPC, or owner funded energy projects.

Even though it appears counter intuitive, often EPC projects are not limited to energy conservation or energy capacity enhancement projects, instead, the breadth of their scope includes water conservation, sustainable materials and operations. Examples of some of the mainstream technical measures, typically included in the EPC projects, include the following:

Lighting

Replacement of inefficient lighting systems with energy efficient lamps, energy efficient ballasts and optimally designed light fixtures. Examples include:

- Replacement of mercury lamps with higher efficacy sodium vapor lamps
- Replacement of incandescent and, in some cases, florescent lamps with high efficacy LED, light emitting diode, lamps
- Replacement of older florescent lamps and fixtures with high efficiency T-8 or T-12 florescent light systems.
- Replacement of lighting systems, that are inadequately designed by today's standards, with light systems that are designed with emphasis on important factors such as lighting efficacy (lumens/watt), CU, coefficient of utilization (lumens reaching the work plane/total lumens generated).

HVAC, Heating, Air Conditioning and Ventilation

Optimization and improvement of chilled water systems, replacement of lower efficiency and high maintenance HVAC Systems with HVAC systems that carry high Energy Star rating, utilize high efficiency chillers, use green technologies, i.e. geothermal, solar, thermal storage, etc.

Control Systems

Control systems incorporating effectively designed and optimally applied control architecture in energy usage and energy generating systems. These control systems employ cutting edge—yet proven—sensors, transducers and other control devices for field application. Furthermore, these control systems are driven by CPUs, central processing units, or computers and PLCs, programmable logic controllers that offer the latest improvements in hardware, firmware, application software and HMI, human machine interface, options.

Building Envelope Improvements

Measures in this category include infrastructure improvements related to the building envelope or exterior. Examples of building envelope improvement are:

- Refurbishment of roof systems
- Replacement or upgrade of insulation
- Installation of energy efficient glazing (windows)

Cogeneration and CHP

Cogeneration and CHP, combined heat and power, measures address the supply side of the energy equation. Measures in this category can include the following:

- Topping cycle cogeneration system
- Bottoming cycle cogeneration system
- Combined cycle cogeneration system

Combined cycle systems employ both the topping cycle feature as well as the bottoming cycle feature and, therefore, offer the higher efficiency.

Demand Response Measures

Demand response measures are projects or actions undertaken to avert the need for power generating capacity expansion. Demand response measures are also referred to as DSM, or demand-side management, measures. DSM is an important tool to help balance supply and demand in electricity markets, to reduce price volatility, to increase system reliability and security. This enables the utility industry to rationalize investment in electricity supply infrastructure and to reduce greenhouse gas emissions. Examples of these measures include the following:

- Energy efficiency enhancement technologies, management practices or other strategies in residential, commercial, institutional or governmental arena that reduce electricity consumption.
- Demand response or load management technologies, management practices or other strategies in residential, commercial, industrial, institutional and governmental arena that shift electric load from periods of peak demand to periods of off-peak demand, including pump storage technologies.
- Industrial by-product technologies consisting of the use of a by-product from an industrial process, including the reuse of energy from ex-

haust gases, steam or other manufacturing by-products that are used in the direct production of electricity.

Figure 11-1 is Quantity (Q) – Price (P) graph. This graph shows the effect of demand response on demand elasticity. The inelastic demand in the electrical power market place is represented by curve D1. The high price P1 associated with the inelastic demand D1 is extrapolated off the point of intersection of the supply curve S and the demand curve D1. When demand response measures are introduced, demand becomes elastic. The elastic demand is represented by curve D2. The point of intersection of elastic demand curve D2 and the supply curve S precipitates in a substantially reduced market price P2.

It is estimated that a 5% lowering of the demand would result in a 50% reduction in price during peak hours, as demonstrated in the California Electricity Crisis of 2000-2001.*

Other studies, such as the two studies sponsored by Carnegie Mellon in 2006**, examined the impact of demand response measures on demand

*The Power to Choose—Enhancing Demand Response in Liberalised Electricity Markets Findings of IEA Demand Response Project, Presentation 2003.
**CEIC-07-01 "Demand Response and Electricity Market Efficiency," Kathleen Spees and Lester Lave. CEIC-07-02 "Impacts of Responsive Load in PJM: Load Shifting and Real Time Pricing" Kathleen Spees and Lester Lave

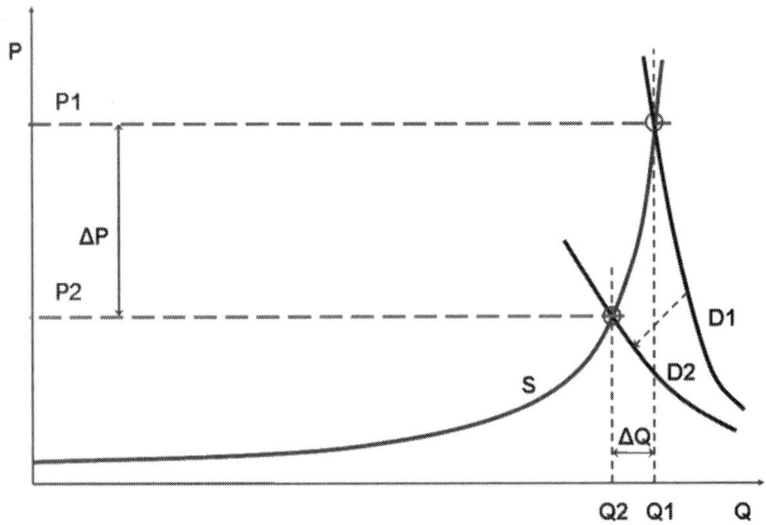

Figure 11-1. Demand elasticity and the effect of demand response

elasticity and price. The price reduction can be explained by the fact that operators generally plan to use the least expensive, or lowest marginal cost, generating capacity first, and use additional capacity from more expensive plants as demand increases.

Biomass Source Measures

Biomass is a relatively clean renewable energy source that can assist in diversification of transportation fuels. Figure 11-2 shows the product and process flow associated with biomass fuels. Some of the challenges and contributors to the cost side of the equation are:

- R&D focus on developing and optimizing cost-effective, integrated systems for collecting, storing, preprocessing feedstock
- Feedstock logistics; transporting a range of cellulosic feedstocks, including agricultural residues, forest resources and dedicated energy crops.

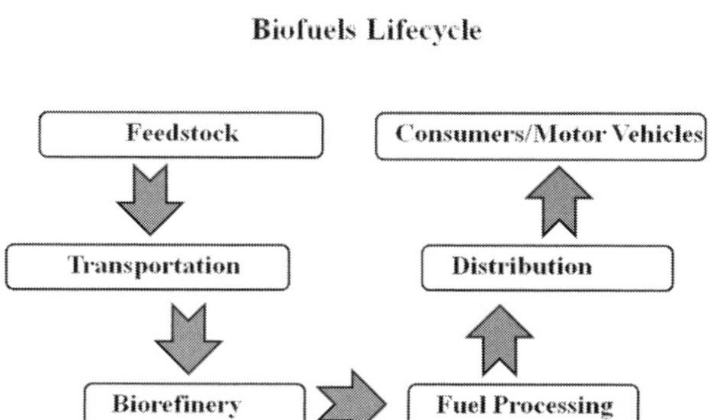

Figure 11-2. Typical Biofuel Lifecycle

Renewable Energy Measures

Renewable is energy that comes from natural resources such as sunlight, wind, rain, tides, and geothermal heat; these are energy sources, or forms of energy, that are renewable and replenished naturally. Data published in 2006 showed that 18% of the total global energy is derived from renewable sources with the breakdown as follows:

- Traditional biomass, primarily for heating: 13%
- Hydroelectric power generation: 3%

- New renewables, i.e. small hydro, modern biomass, wind, solar, and geothermal, while growing rapidly, in 2006, accounted for 2.4%

Renewables account for 18% in the realm of electric power generation, distributed as follows:

- Hydroelectric: 15%
- New renewables: 3.4%

Water and Sewer

Consumption of water and discharge into sewage systems, at face value, may not appear to have a direct relationship with the energy. However, as one examines the process behind the reclamation and distribution of potable water, the vital and indispensible role that energy plays becomes palpable. It takes energy, and therefore dollars, to drive the pumps to collect untreated water. The filtration and sanitization phases of water purification require hydraulic head to move water, which amounts to conversion of electrical energy to hydraulic head. In addition, significant and continuous amount of electrical energy is needed to store and pump water to consumers at a standard head. Hence, it's obvious that if the consumption of water is reduced, energy is conserved and the demand for electrical energy abates.

Sustainable Materials and Associated Operations

Mining, transportation, refinement, treatment, sizing, packaging, marketing, sale, distribution and application of all types of materials—pure metals, alloys, non-metallic substances, polymers and myriad synthetic substances—require energy in all phases of production. Even recycling of materials, as "green" as it is, requires energy. As we assume this perspective, it becomes easy to see how conservation in the use of materials can have an impact on energy demand.

ESCO stands for energy service company. One way to understand the relationship and distinction between EPC and ESCOs is to think of EPC, energy performance contracting, as a process and the ESCOs as entities that implement the EPC process. ESCOs can provide a full range of services required to complete an energy project. Such services, often, include the following or a combination, thereof:

- Energy audit
- Design engineering

- Construction management and supervision
- Facilitation or provision of project financing
- Start-up and commissioning
- Operations and maintenance
- Monitoring and verification of energy savings

Historically, the inception of EPC could be traced as far back as the early eighties. In the pre-1985 era, ESCOs were established, as a part of the DSM, demand-side management, efforts to provide personnel and equipment resources to the utilities as they strived to meet the energy conservation mandates imposed by federal and state governments.

From the mid-1980s through 2003, ESCOs, and EPC industry as a whole, have seen substantial ebb and flow in growth, acceptability and revenue. Over this period, some of the ESCOs have transformed and some have grown either organically or accretively, through consolidation. This evolution within the EPC domain was influenced, favorably, by the state and federal governments. The successes of the EPC industry, in the 1994-2002 period could, to a certain extent, be attributed, to studies by LBNL, Lawrence Berkley National Laboratory, and NAESCO, National Association of Energy Service Companies, that highlighted the EPC successes and encouraged the state and federal governments to promote EPC. Another important event that could be credited for EPC growth and successes in the 1994-2002 period was the formulation and implementation of the IPMVP, or International Performance Measurement and Verification Protocol. The IPMVP provided standard methods for documenting project savings and provide commercial lenders the confidence to finance EPC projects.

As plausible, the EPC industry was impacted unfavorably by the ENRON debacle, as ENRON was a significant player in the ESCO market. The ENRON collapse coupled with the uncertainty about the deregulation of the electric utility industry impeded the growth of EPC in the 2002 to 2004 period.

The 2004-2006 period showed 20% growth in the EPC industry with comparable projected growth trend. The 20% growth, and subsequent upward trend, can be attributed to volatility in the energy market and the increasing energy prices. Other contributors to the heightened interest in the EPC are state and federal mandates, inadequate capital and maintenance budgets for federal and state facilities. Growing awareness of the greenhouse gas emissions and realization of the fact that large-scale, sustained, remediation is needed in this regard has made EPC more attractive at local, state and federal levels, and in the private sector.

While ESCOs, in response to customer requests, are constantly adding new measures and services to their project portfolios, they are not to be construed as stewards for technological research, development and marketing in the energy domain. ESCOs and their clients tend to be fairly conservative and risk averse in selection of technologies for projects. Due to the fact that the cost of most ESCO projects are paid from energy savings, often secured with financial guarantees, ESCOs and their clients tend to lean in favor of proven technologies.

INDUSTRY REVENUE SEGMENTATION BY PROJECT TYPE OR MEASURES UNDERTAKEN

There is frequent reference to the terms "project" and "measures" throughout this text. These two terms are synonymous and represent discrete effort, consisting of labor and material, expended to affect energy conservation, energy productivity or energy cost reduction. In the beginning of this chapter we discussed the types of projects or measures that are offered and implemented by ESCOs, within the EPC domain. Then, in the previous section we noted that EPC industry and ESCOs provide various services to their clients—related to the EPC projects—ranging from energy audits, design engineering, all the way through commissioning, measurement and verification. Now, let's explore the overall EPC ESCO market landscape further by analyzing the market on the basis of the technical essence of the projects or measures undertaken. Figure 11-3 shows ESCO industry revenue segmentation by general project or measure type. This chart, similar to some of the other statistical information presented in this chapter for relative comparison purposes, is based on 2006 LBNL and NAESCO studies. As shown in Figure 11-3, energy efficiency type projects constituted lion's share of all projects. The energy efficiency measures amounted to approximately 73% of all EPC projects implemented. Onsite renewables and onsite turbine/engine based power generation represented a combined share of about 16%. Eight percent (8%) of the EPC revenue transactions were associated with relative intangible measures such as consulting, planning, management and coordination. Project implementation related, indirect, overhead and administrative costs would be a portion within the 8% revenue segment. The fact that

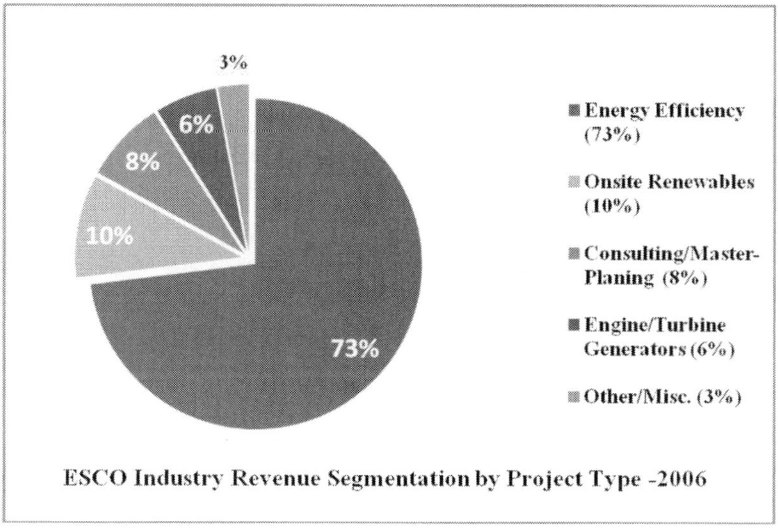

Figure 11-3. ESCO Industry Revenue Segmentation by Project Type or Measures

only a small fraction of the revenue pertains to overhead and intangible administrative activities attests to the overall cost effectiveness and productivity of EPC approach.

ESCO MARKET SEGMENT COMPARISON ON THE BASIS OF SEGMENT REVENUE

The markets served by ESCOs can be better understood by examining the market segmentation chart depicted in Figure 11-4. This chart divides the overall EPC ESCO market into five major segments based on revenues. These market segments are as follows:

- **The MUSH Market Segment**: The MUSH—municipal, university, school and hospital—market segment includes municipal and state governments, K-12 schools, universities, colleges and hospitals. The MUSH markets, as shown in Figure 11-4, represent a dominant share of the overall ESCO market, at approximately 58%. The MUSH market share has held quite steady over past several years.

- **The Federal Market Segment**: The federal market segment constitutes, approximately, 22% of the overall ESCO EPC market. Even though the federal market segment does not hold a dominant share of the overall ESCO market, their importance and their influence has become quite prominent in that last 10 years.* ESCOs offer energy services to federal agencies through various financial mechanisms. ESCOs not only provide EPC services to federal agencies but also offer design and build function, supervise contracts and act as contractors on UESC, utility energy services contracts. Based on the OMB, Federal Office of Management and Budget, the energy efficiency investment in federal facilities approached $668 million (Vallina 2007).

- **Industrial Segment**: Approximately, 6% of the ESCO revenue is derived from the industrial market segment. The small size of this market segment is due to the fact that private industry prefers to invest *capital* on projects that yield productivity gains, safety improvement, reduced labor cost and enhanced sales.

- **Commercial Segment**: Approximately, 9% of the ESCO revenue is derived from the industrial market segment.

- **Residential Segment**: Approximately, 3% of the ESCO revenue is derived from the industrial market segment.

- **Public Housing**: Approximately, 2% of the ESCO revenue is derived from the industrial market segment.

While ESCOs, like other business organizations and institutions, continue to evolve and change to a certain degree, they can be categorized into the following four, major, categories:

- Independent ESCOs
- Utility ESCOs
- Building Equipment, or simply, Equipment Manufacturer ESCOs
- Other Energy/Engineering Services ESCOs

*See Hopper et al. (2005) for discussion on procurement approach that has boosted the growth of the federal market segment.

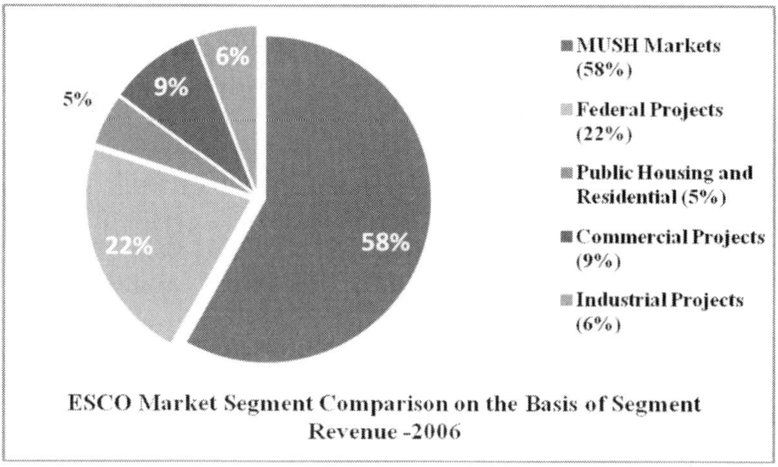

Figure 11-4. ESCO Market Segments Based on Revenue

Independent ESCOs

Independent ESCOs are energy service firms that are not connected with utility companies or equipment manufacturers. Many independent ESCOs concentrate on specific customer market segments and tend to focus their efforts on select geographic regional markets. The fact that independent ESCOs are not associated with any specific equipment manufacturer is touted by these ESCOs as a marketing strength or advantage. This claim is premised on the notion that, since the independent ESCOs have no alliance with specific equipment brands, they are more likely to apply equipment that is most suitable for a given energy project without brand loyalty constraints.

Utility ESCOs

Utility ESCOs are owned by electric or gas utility companies. They tend to target markets in the same regions as their parent utility companies. Also, as expected, they benefit from the sponsorship of their parent utilities in the markets served by those utility companies. These attributes could be considered as "strengths" of the utility ESCOs in the regional or local geographic markets. The name recognition and customer relationship, enjoyed by the utility ESCOs in the local and regional markets, could be considered marketing and business strengths, and could be assigned as competitive advantages for the utility ESCO.

Building and Control Equipment Manufacturer ESCOs

As the name implies, these ESCOs are owned by, or are subsidiaries of, major building and controls equipment manufacturers. Since most building equipment and controls manufacturers have national, and in some cases international, presence, these ESCOs have an extensive network of branch offices. Building and control equipment ESCOs are, generally, large in territorial respect and tend to have the capital resources needed to compete in markets that are capital intensive and where project costs are high. The building and control equipment ESCOs also tend to hold a competitive advantage over other ESCOs from brand recognition and brand loyalty perspective. In that, a client or customer that has a national agreement with Brand "X" controls equipment manufacturer/supplier will tend to favor this brand on EPC projects. The reasons for such preference include, level of confidence and trust in the brand, commonality of spare parts, minimum maintenance training cost and preferential service due to volume of business.

Energy and Engineering Services ESCOs

These ESCOs are owned by, or are subsidiaries of, major oil or gas companies, non-regulated energy suppliers or large engineering firms. The strategic strength or competitive advantage of engineering service type ESCOs lies in the fact that they hold sizeable in-house engineering expertise as compared to the other ESCOs. This could provide a technological advantage to engineering service ESCOs on EPC projects that are custom and require significant amount of engineering.

MARKETING AND BUSINESS PERSPECTIVE, EPC AND ESCOs

Figures 11-3 through 11-7 are based on market, revenue and other business data collected by LBNL and NAESCO over 2000-2006 time span. Since the EPC industry, as many others, is dynamic and in a state of flux, the value of the data presented lies in drawing a relative comparison and should not be construed in the absolute sense.

Figure 11-5 compares the size of the four major ESCO categories based on the number of players or ESCOs within each category. As apparent, the engineering services ESCOs occupy the least market share, at approximately 10%. On the other hand, the independent ESCOs carry the largest market share at, approximately, 61%. The equipment and utility ESCOs fall in the

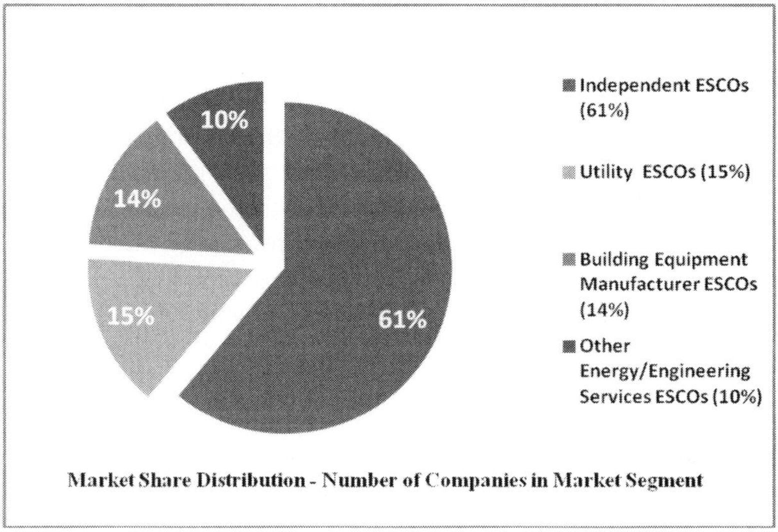

Figure 11-5. Market Share Distribution—Number of ESCO Companies in Market Segment

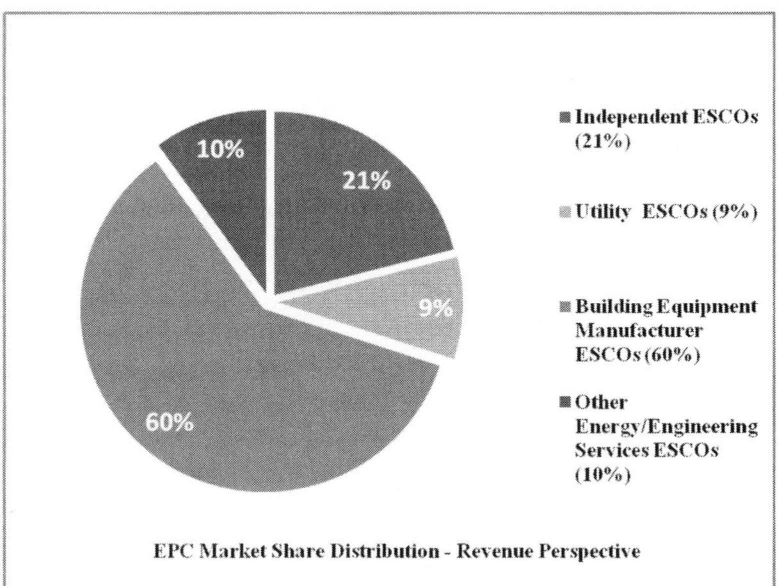

Figure 11-6. EPC Market Share Distribution—By Magnitude of Revenue

middle, with an almost equal share. The reason for independent ESCOs representing the largest volume is that independent firms within the independent ESCO category are numerous. However, the firms that the independent ESCO category is composed of tend to have smaller revenues. The later observation is illustrated through Figure 11-6 where market share of each of the four ESCO categories is examined on the basis of revenue.

Figure 11-6 illustrates the dominance of the building equipment and controls manufacturer ESCOs with an approximate 60% market share, on the basis of revenue. It appears that the geographic dominance, technological strength, greater access to investment funds and brand loyalty tend to serve as formidable strategic advantages for the equipment manufacturer ESCOs.

As shown in Figure 11-6, the utility ESCOs have the smallest market share on the basis of revenue, at 9%. Their market share in 2000 was about 39%. This weak and drastically curtailed market position of the utility ESCOs, from revenue perspective, can be explained on the basis the fact that the parent utility companies have diverted their strategic focus to business segments that offer quicker returns. Pursuant to this shift in their business strategy focus, the parent utility firms have divested, spun off, or closed most of their ESCO businesses. This significant reduction of Utility ESCO market share also underscores the fact that business strengths that rely heavily on local or regional markets tend to be volatile and less enduring.

Any business analysis would be incomplete without analysis and evaluation of the strategic strengths, weakness and competitive advantages. Figures 11-7 and 11-8 help us do just that for the EPC ESCO industry.

Figure 11-7 compares the four major ESCO segments on the basis of **financial/capital strength** and **geographic presence** or range. Each of the four players is placed on the matrix shown in Figure 11-7 based on the range of their geographic presence and their, relative, capital or cash strength. The building and control equipment manufacturer ESCOs are located in the top rightmost corner of the matrix because they not only enjoy a widespread, local, regional, national and international presence but, compared to other ESCO segments, they also have the greatest access to investment capital. This business position allows the building and controls equipment ESCOs to enjoy a distinct strategic and competitive advantage over the other ESCOs. This strategically strong position of the building equipment and equipment manufacturer ESCOs, as highlighted in Figure 11-7, comports with their revenue dominance as shown in Figure 11-6.

On the other end of the spectrum—as borne out by examination of the revenue market share analysis—the utility ESCOs have a weaker and strategically less advantaged position. The utility ESCOs are located in the lower midsection of the matrix, signifying the fact that they have the smallest footprint in the overall EPC ESCO market, and due to the significant divestiture by their parent utility firms, they have diminished access to capital for EPC projects.

The strength and competitive advantage matrix depicted in Figure 11-8 places the four ESCOs on the basis of their technological strengths and their range or geographical presence. Once again, the Building and Control Equipment Manufacturer ESCOs tend to surpass the competing EPC ESCO segments in competitive advantage, boosted largely by their technological strength. When examined on the geographical presence and technological advantage plane, the Utility ESCOs are placed on lower left segment of the matrix. They lag behind the other three EPC market players and appear to assume a withdrawn position.

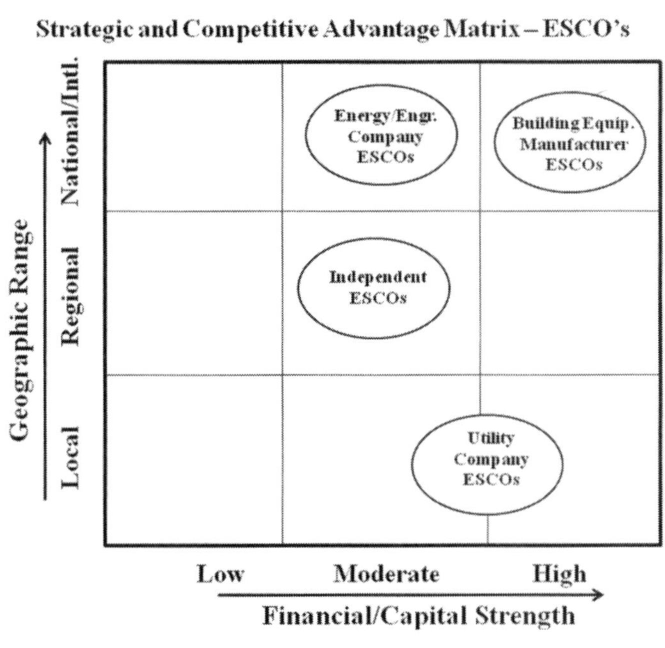

Figure 11-7. ESCOs—Strategic and Competitive Advantage Matrix, Geographic Range Versus Financial/Capital Strength

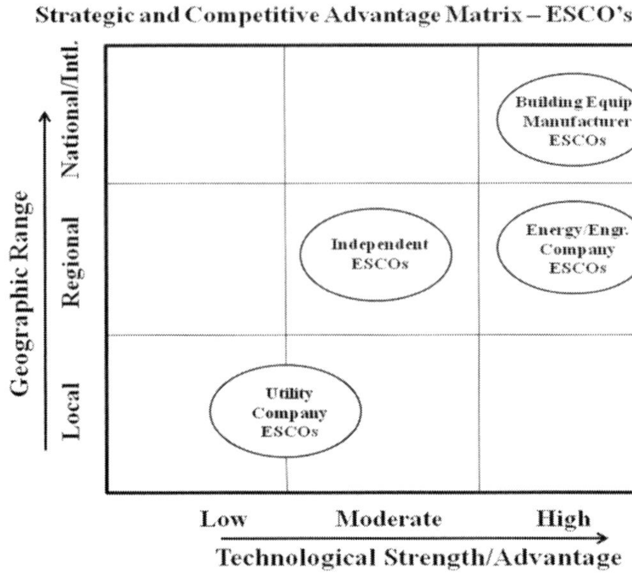

Figure 11-8. ESCOs—Strategic and Competitive Advantage Matrix, Geographic Range Versus Technological Strength

However, the diminished competitive advantage and limited geographic presence of the Utility ESOCs, is by no means a reflection of business strength or weakness of the parent utility companies. Most parent utility companies continue to flourish in energy, power generation and energy distribution domain. Most utility companies continue to maintain their respective business strengths, the loyalty of their customers and confidence of their shareholders, within their respective territories.

EPC FINANCING PERSPECTIVE

Several proven and viable financing alternatives are available for various industrial, commercial, local, state and federal EPC projects. Some financing vehicles appeal more to the industrial and commercial EPC projects while others are more practical and congruent with the local, state and federal government culture. The financing terms, conditions and vehicles chosen vary based on the merits and specific requirements of each project. Large institutional lenders have made billions of dollars available for EPC projects over the last 10 years.

Some of the financial instruments or financial vehicles that have been employed, in recent years, by the EPC project financiers are listed below:

1. **Operating Expense Financed EPCs**: Operating expense financed EPC projects are structured, financed and contracted such that the repayment of the borrowed funds is made through savings in the periodic operating expenses. As explained in Chapter 1, operating expenses include utility bills, direct or indirect variable costs associated with production, fabrication and processing activities. Of course, on operating expense financed EPC project, the savings used to recoup the cost of the project are based, primarily, on energy or utility cost savings. In that respect, the operating expense based financing method is similar to, or is component of, the more commonly utilized financing methods, such as, the tax-exempt lease-purchase agreements.

2. **Capital Expense Financed EPCs**: With the capital expense financed EPC projects, the capital expense budget is used to pay for the long term debt associated with the EPC projects. Such debt can include, but not limited to, acquisition of fixed assets, i.e. buildings, furniture, on-site power generation, chillers, HVAC equipment, lighting, etc. Repayment associated with capital expensed EPC projects extends through the operating period. While capital expense financing approach provides the freedom from protocol and procedures associated with governmental projects, it carries the following disadvantages:

    a. Scarcity of capital; especially, during less than exuberant economic times when reduced sales and revenue impact cash flow unfavorably and tend to depress market confidence.

    b. Competition with other, non-energy, projects is another concern with the use of capital expense means as a financing vehicle for EPC projects. Financially, energy projects tend to bear long payback periods, and relatively unfavorable financial ratios as compared with productivity enhancement or marketing projects. Safety and environment projects tend to trump all other projects, in many cases.

    c. Possible, unfavorable impact on the organization or institution's ROE or ROI. This is due to a direct, and relatively immediate, inclusion of capital expenditures into the asset base.

d. Relatively long funding authorization or approval times. In the commercial or industrial arena, the approval process could span over months due to sign-off by multiple layers of management. For EPC projects in the local, state or federal realm, funding for significant EPC projects may be subject to voter approval. In fact, MUSH or other public entity agreements, involving assumption of debt by such institutions, requires taxpayer approval.

3. **Cash from Capital Budgets**: This method for full or partial financing of EPC ESCO projects simply involves the use of cash allocated from the state or local capital budgets. Stipulations governing the allocation and use of state or local capital funds vary from state to state and from one municipality to another. Some states and local governments prohibit the use of capital funds for EPC projects.

4. **Capital Leases**: Capital lease is a financing vehicle that permits clients to purchase capital equipment on installments. Capital lease method of financing differs from the traditional capital financing method in that the no initial capital is required. The owner or principal assumes ownership of equipment at the end of the lease term. Also, because capital is involved, the owner-lessee may take deduct the interest portion of the lease payments. A capital lease is recorded on the owner-lessee's balance sheet like a financed capital type transaction. The initial cost is recorded as asset on the asset side of the balance sheet "equation."

Assets = Equity + Liability

The amount owed on the project, is recorded on the liability side of the equation. As principal is paid off through the lease payments, it is recorded as equity, thus keeping the asset, equity and liability equation balanced.

On capital lease projects, the interest earned by the lessor, or financing entity, is tax exempt. Therefore, similar to tax-exempt lease-purchase agreements, the lender has an incentive to offer financing at interest rates that are lower than the ones offered for non-tax exempt financing agreements.

A lease must meet the following criteria in order for it to be classified as a capital lease:

a. The value of the lease must meet or exceed 90% of the recognized fair market value of the equipment or assets, at the time of initiation of the lease.

b. The terms of the capital lease agreement must include transfer of ownership of all assets to the lessee.
c. The lease term must span over, at least, 75% of the estimated economics life of the equipment or assets.
d. The lease contract must include a bargain purchase option.

Note: The criteria stated above are FASB, Financial Accounting Standards Board, rules. Government entities, including MUSH, may be eligible for tax-exempt capital lease agreements, and many MUSH entities find the capital leases appealing.

5. **Shared Savings**: The shared savings agreement involves predetermined, contractually agreed upon, distribution of measured savings between the owner and the ESCO/contractor. In the absence of energy savings, the principal entity (owner) is only responsible for the utility bill, and no payment is made to the ESCO/contractor. As with many other EPC ESCO projects, the ESCO bears all of the risks associated with the project:

    a. Project performance risk. If the system or equipment do not deliver the projected or estimated savings, the ESCO or contractor assumes the loss.

    b. Risk associated with fluctuation of interest rates, where the financing is based on variable interest rates.

    c. Risk associated with escalating utility cost rates. Any increase in the energy cost rate has a direct impact on the profit and losses of the ESCO.

With shared savings agreements, ownership of the equipment transfers to the customer in accordance with the shared savings contract. Because of the heavily lopsided risk assumption by the contractor or service provider, shared savings type agreements are not pursued or promoted by the ESCOs. Federal entities are an exception, where the federal mandates require participation in EPC ESCO contracts through shared savings agreements.

6. **Revolving Loan Pools**: Certain states, such as Texas and California, allocate certain amount of funds for EPC type projects. These funds

are used as a revolving loan pools offered at financing rates that are significantly lower than the rates available through private or commercial financial institutions. Since the interest rates through revolving loan pools are quite attractive, there are waiting lists of projects and clients vying for participation in such programs. New loans or projects have to wait for existing projects in the revolving loan pool to be paid off. This can result in years of delay on projects and, therefore, revolving loan pools are not suitable for programs that require implementation within 12 to 24 months.

7. **Power Purchase Agreements**: The PPAs, power purchase agreements, are typically associated with distributed power generation and CHP, combined heat and power measures. Customers participating in the PPAs agree to purchase the output of a facility, projects or measure—in terms of kWh of electrical energy or MCF of steam, at a given pressure. The client or facility owner enters into a contract with the ESCO to purchase the utility at a certain cost rate. If the cost of electricity through the local power company drops below the contract price charged by the ESCO, the facility owner has the option to break the contract. Conversely, if the provision of utilities at contract prices precipitates into losses, for the ESCO, the later has option to dissolve the contract. Facility owners participating in PPAs only pay for the energy or utility consumed. The PPAs type EPC programs are premised on contracts that are the least obligatory in comparison with some of the other ESCO financing programs.

8. **Tax Exempt Lease Purchase Agreements**: This method of financing an EPC ESCO project is one of the most common methods, especially in the MUSH, market. The acronym "MUSH" stands for municipal, university, school and hospital market segment. This EPC project financing vehicle involves repayment of borrowed funds with the savings obtained through operating expense reduction—which, in most EPC projects is due to energy/utility productivity enhancement. As explained earlier in this section, using operating expense savings to pay for EPC projects is more effective than the traditional debt financing tools such as bonds, commercial loans, etc. Some of the more important features of the tax-exempt lease-purchase agreements are as follows:

a. Tax-exempt lease-purchase agreements are sometimes referred to as municipal leases. Tax-exempt lease-purchase agreements, unlike the traditional, commercial, or personal equipment leases, transfer the ownership of the assets to the lease holder at the end of the lease term. This can be especially attractive when, as in most EPC ESCO projects, the life of the equipment far exceeds the term of the term of the lease agreement.

b. The effective interest rates associated with financing of the tax-exempt lease-purchase agreements based EPC contracts are significantly lower that the ones offered on taxable commercial lease-purchase agreements. This is due to that fact that the interest paid on such debt instruments by public entities (i.e. the institutions in the MUSH market) to the financial institutions are exempt from federal income tax. This exemption from federal income tax serves as an incentive for financial institutions to lower their interest rate for the EPC ESCOs or the owners.

c. Customers, clients or owners have to demonstrate and assure and lending institutions that the energy efficiency projects are an essential part of their operations, the energy efficiency measures will be utilized and the energy or utility savings will be realized. This boosts the lenders confidence and minimizes the non-appropriations risk. As expected, the lower the risk—or in some cases, perceived risk—the lower the interest rate and more favorable the terms on EPC project financing.

d. The non-appropriation clause, as explained in Chapter 1, limits the institution's (customer) liability to the ***current*** operating budget period. In that, if future funds are ***not appropriated*** for implementation of the entire plan or project, the equipment may be returned to the lender, or the ESCO, and the obligation is terminated at the end of the current operating period. Since the "non-appropriation" clause is typically included in the tax-exempt lease-purchase EPC agreements, this type of financing may be considered to constitute an operating expense rather than capital expense. And, when MUSH, or other public entities, enter into agreements involving expense, taxpayer approval, in most states, is not required.

It is noteworthy to mention that the net benefit of tax-exempt lease-purchase EPC approach may vary from state to state, county to county, even between one municipality and another, based on the governing statutes. The terms, legal constraints and stipulations that affect the net impact of a tax-exempt lease-purchase may differ between one school district and another within the same county.

Therefore, it is advisable that qualified and experienced legal counsel be engaged in tax-exempt lease-purchase EPC projects. Other caveats that must be noted in this regard are as follows:

a. Approval process associated with the tax-exempt lease-purchase EPC project is arduous under certain circumstances.
b. Alternate funding sources of funds may be accessible and readily available with more favorable terms, through other financial instruments such as, governmental bonds.
c. Some state and municipal statutes or charters may not permit MUSH or other institutions to enter into tax-exempt lease-purchase agreements.

9. **State or Local Government Leasing Pools**: The state or local government pools are a means for financing energy or utility conservation measures through a financial vehicle called COPs, or certificates of participation. The states or local governments, providing leasing pools, build a collection of funds by selling COPs pertaining to energy or utility type projects. This pool of funds is used to finance tax exempt EPC ESCO lease agreements for the state and local government agencies. The COPs constitute a legal document certifying that the holder of the document is a partial owner of an EPC ESCO lease-purchase agreement. COPs are somewhat like bonds, in that the investors purchase them as financial investments. However, the COPs do not offer the liquidity and security that the regular financial bonds do. The COPs could serve to provide an investor a means for diversifying an energy/utility type financial investment portfolio.

10. **State or Local Government Bonds**: State and local government bonds are a simple way to finance EPC ESCO projects. State or local governments simply issue bonds to fund EPC type energy projects. In most cases, the state governments will issue a bond to finance multiple projects. This minimizes cost of issuing the bonds and the interest rate offered to the ESCOs or owners. As with most project finan-

cial pooling, financing of energy projects through state government bonds is not expedient and involves wait time associated with collection of projects before bonds are issued and funds made available for the energy/utility projects. On the other hand, the local bonds tend to issued for specific EPC ESCO projects. Since there is no pooling involved, the wait times for local government bonds is, relatively, short. However, the local bonds, in most cases, have to be approved by the local government's legislative body, such as the City Council. In some cases, the local bonds are subject to approval by voters.

EPC MEASUREMENT AND VERIFICATION CONSIDERATION

Measurement and verification, abbreviated as M&V, is not only crucial for control of energy or utility productivity but also essential for tracking cost savings associated with EPC ESCO projects. Verification of actual savings requires measurement and verification Therefore, the design and implementation of M&V, architecture, hardware, application software and protocol are important elements in the success of an EPC ESCO project. Early EPC ESCO projects relied on energy or utility savings based on projections—supported by laboratory tests and statistical data. Measurement and verification of energy and utility savings on EPC projects can be conducted on real-time basis. These measurements can be made through deployment of state of the art sensors, transducers, meters, gages and other pertinent instruments. In many systems, the accuracy of M&V systems ranges from 0.5% to 1%. Contracts, such as the Shared Savings EPC ESCOs, depend on verified savings. This led to the development of IPMVP, International Performance Monitoring and Verification Protocol*. This protocol was developed by a joint collaboration between ASHRAE, U.S. DOE, and NAESCO.

Notwithstanding the M&V alternatives discussed above, the financiers of projects, the EPC ESCOs and even the facility owners prefer to rely on stipulated or projected savings. Some facility owners or customers resist substantial investment in M&V technology; they prefer to maximize capital investment in actual energy productivity enhancement measures. The financiers prefer the stipulated savings approach for their return on

*Introduction to Energy Performance Contracting; U.S. Environmental Protection Agency, Energy Star Buildings; IFC International and National Association of Energy Services Companies; October 2007.

investment due to the preception that the stipulated or projected savings approach minimizes the uncertainty of payments associated with the measurement and verification. In other words, if payments to the financiers and ESCOs are based on the M&V based savings, no payments are made by the facility owner unless savings are actually measured and verified. Efforts are now underway to develop protocols that utilize state of the art sensing and measurement technology, capture actual energy and environmental improvements, and promote the development of more accurate stipulations for EPC ESCO contracts.

CASE STUDIES

Case Study 11-1. Demand-side Management, Demand Elasticity and Electricity Price Reduction:

A large industrial customer in California is considering participation in a DSM (demand-side management) or demand response project. The current, average, energy cost rate for this large industrial consumer is $0.13/kWh during peak hours. This facility operates on a 24/7 schedule. For simplicity, assume that the peak hours stay constant, at 8 peak hours per day.

(a) Based on the demand elasticity model depicted in Figure 11-9, estimate annual electrical energy cost reduction that could, potentially, be realized though this joint collaboration between the industrial consumer and the utility if a 7% peak demand reduction is achieved through Demand Response Measures. The cost savings are to be based on the existing demand of 30 MW over a 30-day billing period. Use 365 days for annual cost calculation. Assume that the California Electricity Crisis model of 2000-2001, as discussed earlier in this chapter, applies.

(b) What would the cost reduction be, in percent (%), due to the 7% on-peak demand reduction?

Solution:

The solution strategy for this case study is centered on the demand elasticity model shown in Figure 11-9 and the California Electricity Crisis model of 2000-2001.

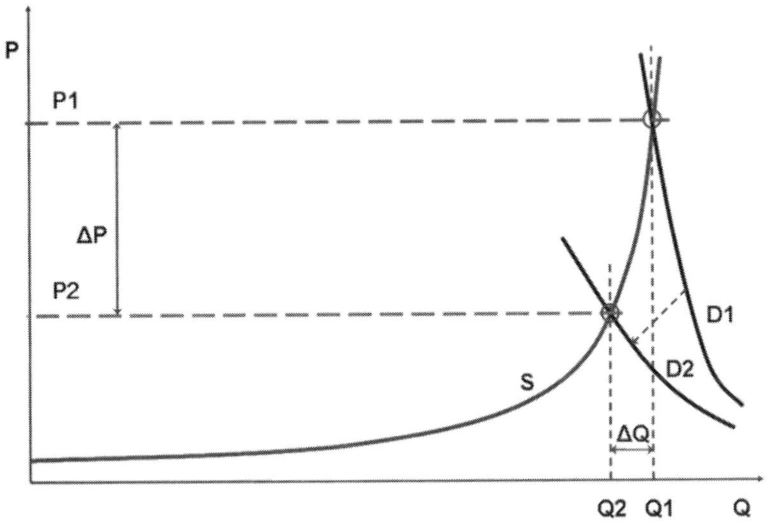

Figure 11-9. Demand elasticity and the effect of demand response

Given:

- P_1, the current peak demand cost rate for the industrial customer is \$0.13/kWh
- Q_1, Current demand: 30 MW
- Number of peak hours per day: 8
- Demand reduction through demand response measures: 7%
- California Electricity Crisis model of 2000-20011 applies. Therefore, an estimated 5% lowering of the demand could result in a 50% reduction in price during peak hours.

Part (a): Based on the demand elasticity model depicted in Figure 11-9, estimate annual electrical energy cost reduction that could, potentially, be realized though this joint collaboration between the industrial consumer and the utility if a 7% peak demand reduction is achieved through Demand Response Measures.

Current On-peak Energy Usage per year, in kWh

= (30 MW) x (1,000,000 W/MW) x (1 kW/1000 W) x (8 hours/day) x (365 days/year)

= 87,600,000 kWh

Current On-peak Energy Cost per year
= (87,600,000 kWh) x ($0.13/kWh)
= $11,388,000

Projected On-peak Demand, in MW, after implementation of the Demand Response Project
= Q_2
= (30 MW) x (1 – 0.07)
= 27.9 MW

Projected On-peak Energy Usage per year, in kWh
= (27.9 MW) x (1,000,000 W/MW) x (1 kW/1000 W) x (8 hours/day) x (365 days/year)
= 81,468,000 kWh

Projected % Reduction in energy cost rate due to 7% demand reduction, through demand response measures, based on California Electricity Crisis model of 2000-2001.

= P (%)
= (50% Price Reduction/5% Demand Reduction) x (7% Demand Reduction in this Case Study)
= 70%

Projected On-peak Energy Cost Rate
= P_2
= (1 – 0.7) x ($0.13/kWh)
= $0.039/kWh

Projected On-peak Energy Cost per year
= (81,468,000 kWh) x ($0.039/kWh)
= $3,177,252

Therefore, the projected annual energy cost reduction due to 7% demand reduction, through demand response measures:

= $11,388,000 – $3,177,252
= $8,210,748

Part (b): What would the annual cost reduction be, in percent (%), due to the 7% on-peak demand reduction?

Annual Cost Reduction, in %
= ($11,388,000 – $3,177,252)/($11,388,000) x 100%
= 72%

Ancillary Discussion

The use of California Electricity Crisis model of 2000-2001 results in a rather high, 72%, projected reduction in the cost of on-peak electrical energy due to demand response based 7% reduction in peak demand, an almost 10-fold effect. While the use of the California model in this case study is illustrative, it should noted that:

a) Many similar, historical models—involving lower energy cost regions of the US—show energy cost reductions to the tune of three (3) to four (4) fold.

b In most cases the larger magnitudes of energy cost reduction are likely to be realized in the wholesale segment of the energy market.

INTRODUCTION TO CASE STUDIES 11-2 AND 11-3

Case studies 11-2 and 11-3 are intended to illustrate the financial principles, strategies and analysis tools that can be employed to evaluate various project financing alternatives. Even though these two case studies compare only a few EPC project financing approaches, the process of comparison can be extrapolated to other approaches with minimal modification.

The analyses in both case studies are based on the same energy/utility efficiency project within the MUSH market. Due to the age of the infrastructure, inefficient utility systems, deterioration of the building envelope, escalating maintenance costs, obsolescence of certain HVAC equipment and due to recently imposed state mandates, HAR County School System is developing plans to implement energy and utility efficiency measures. The contract is expected to be awarded, as a turnkey project, to one successful EPC ESCO firm. However, before awarding of the contract, the school system wishes to examine various approaches available for implementation of the project. The useful life of all equipment included in the measures is 10 years.

The energy and utility efficiency measures to be included in the contract are as follows:

1. Lighting system upgrade to energy efficient lamps and fixtures. ***Short payback.***
2. Installation of VFDs, variable frequency drives, at existing chilled water pumps as energy efficient means for controlling chilled water flow. ***Short payback.***
3. Installation of an automated HVAC control system and an EMS, Energy Management System. ***Short payback.***
4. Replacement of one of the high maintenance, low efficiency, cooling towers with an energy efficient, low maintenance cooling tower; equipped with VFDs and energy efficient fan systems. ***Short payback.***
5. A 10-year, parts and labor, maintenance contract; included in the cost of each line item. ***Short payback.***
6. Replacement of a chiller with a high efficiency unit. ***Long payback.***
7. Modification and expansion of an existing chilled water system. ***Long payback.***

The prevailing discount rate is 8%. For simplicity, assume that the project is commissioned into service at the beginning of Year 1. The operations effected by these measures operate on 24/7 basis and 365 days a year. Total write-off, at current book value, on all equipment being replaced is $50,000. The costs and savings associated with the measures are listed in Table 11-1.

Table 11-1. Costs and Savings Associated with Project Measures

No.	Description	Measure Cost	Annual Savings
1	Lighting System Upgrade	$900,000	$200,000
2	Installation of VFDs	$100,000	$25,000
3	Automated HVAC Control & EMS Systems	$400,000	$40,000
4	Replacement of Cooling Tower	$100,000	$10,000
5	Replacement of a chiller	$300,000	$40,000
6	Expansion of Existing Chilled Water System.	$300,000	$35,000

CASE STUDY 11-2

I. Compare the following two approaches for implementation of the HAR County Energy/Utility Efficiency Project:

A. Loan-based approach for implementation of the project

B. Simple, capital budget funded, investment approach

The financial analysis approach involves the use of project cost and savings data, as laid out in Table 11-1, to develop a cash flow spread sheet. The cash flows are then used to calculate the following five (5) conventional financial indicators that portend the financial strengths and weaknesses associated with each approach:

i. NPV, Net Present Value, based on manual method
ii. NPV based on Microsoft ® Excel financial function
iii. IRR, Internal Rate of Return
iv. ROI, Return on Investment
v. Payback Period, in years

FINANCIAL ANALYSIS OF INVESTMENT ON ENERGY/UTILITY PROJECT WITHOUT EPC ESCO APPROACH—WITH LOAN

I-A. Project Financing Through Cash Borrowed from Capital Budget; Charged to the Operations at the Discount Rate.

Assumed Annual Discount (Interest) Rate: 8%

Annual Payment Toward Amortization of Borrowed Capital
= $2,100,000 (A/P,8%,10) = ($2,100,000). (0.1490)
= $ 312,900.00

Initial Cost of Project: $2,100,000

Project or Equipment Life = 10 Years

All available financial data on the project is sorted and formatted into Spreadsheet 11-1. This spreadsheet shows cash flows for years 1, 3, 5, 7 and 10. Comprehensive financial analysis over the entire 10-year period is shown in Appendix B. The $50,000 write-off of equipment being replaced is captured, in entirety, as negative cash flow in Year 1. The annual payments servicing the $2,100,000 debt include the principal and interest on the loan; amounting to a negative cash of $312,900 per year, over the 10-year life of the project. The 10-year "life" of the project implies that all equipment, software, firmware and overall method of operation would be functional for a period of 10 years.

For simplicity, all maintenance—including software and firmware updates—is assumed to be included in the overall cost of the project. Amounts in the main cash flow segment of Spreadsheet 11-1 are in thousands of dollars. The comprehensive or full version of Spreadsheet 11-1, as depicted in Appendix B, shows all amounts in full format. The annual savings realized by this project—including the cost savings in the energy/utilities and maintenance realms—in the amount of $350,000 per year—are shown as positive cash flows for each year. The combined negative and positive cash flows, for each year, are reflected in the row titled "Net Annual Cash Flows." The next row in Spreadsheet 11-1, titled "Annual Present Values of Cash Flows" lists the calculated present values of each of the net annual cash flows using the following formula:

$$P = F \, . \, (P/F, i\%, n), \text{ or}$$
$$P = (\text{Net Annual Cash Flow for a } n^{th} \text{ Year}) \, . \, (P/F, 8\%, n)$$

The present values for each of the 10 years are added together to yield the NPV, net present value, of the project. The NPV and other calculated financial indicators for the "loan" approach to the project, such as the IRR, ROI and the payback period, are shown in the second part of Spreadsheet 11-1.

The IRR, Internal rate of return, is calculated by taking the average of all net annual cash flows, for the 10 years, and dividing the result by the total investment.

$$IRR = (\$32,100)/(\$2,100,000) = 2\%$$

The ROI, or simple return on investment, is calculated by dividing the net income, or net cash flow—primarily, for a year subsequent to the first year, or average net income from years beyond the first year—by the total investment, or incremental asset.

$$ROI = (\$37,100)/(\$2,100,000) = 2\%$$

The payback period is calculated by dividing the total investment by the net income, or net cash flow—primarily, for a year subsequent to the first year, or average net income from years beyond the first year.

Payback Period = (\$2,100,000)/(\$37,100) = 56.6 years

Spreadsheet 11-1. Financial Analysis of Investment on Energy/Utility Project Without EPC ESCO Approach—Project Financed Through a Loan. Scenario I-A.

Description	Year 1	Year 3	Year 5	Year 7	Year 10
	In \$1,000s				
Costs					
Write-off of Existing Investment,	(\$50)				
Annual Payments on Capital Loan	(\$312.9)	(\$312.9)	(\$312.9)	(\$312.9)	(\$312.9)
Savings					
Annual Savings	\$350	\$350	\$350	\$350	\$350
Net Annual Cash Flows	(\$12.9)	\$37.1	\$37.1	\$37.1	\$37.1
Present Values of Annual Cash Flows	(\$11.944)	\$29.451	\$25.250	\$21.647	\$17.184

Financial Analyses Results	
NPV	\$202,648
IRR	2%
ROI	2%
Payback	56.6 Years

As we examine the financial analysis and the conventional financial indicators for the, non-ESCO, loan method for financing the project, we observe that, with the exception of the NPV, this approach would fall short of most financial approval thresholds in the corporate and the government domains. This can be attributed to the following reasons:

a) All of the required investment for this project is sourced through a loan, with no investment from the institution's capital budget. The positive impact of the use of capital budget funds on the financial ratios is demonstrated in the next segment of this case study.

b) A higher financing interest rate (8%) was used in the computation of the loan payments. If this project is implemented through a tax-exempt lease purchase approach, as shown later in this case study, the interest rate would be lower.

c) Since this scenario does not involve an ESCO, the cost of design, engineering, equipment, software, construction, installation and maintenance is higher than what would be expected through an ESCO, especially a building or controls equipment type ESCO.

Through earlier discussion in this text on the subject of NPV, we learned that a project or proposal with a negative NPV does not present a sound business alternative and must be rejected. Financial analyses in Spreadsheet 11-1 show that the NPV for undertaking the proposed project through a loan is positive $202,648. So, even though the conventional ratios for alternative I-A are not stellar—and these ratios, in fact, make this approach less appealing than the other funding scenarios—from an NPV point of view, the loan approach could present a viable alternative.

I-B. Financial Analysis Of Investment On Energy/Utility Project Without EPC ESCO Approach—Financed, Entirely, Through Capital Budget, With No Loan Involved

Assumed Annual Discount (Interest) Rate: 8%

Initial Cost of Project: $2,100,000

Project or Equipment Life = 10 Years

As with segment I-A in Case Study 11-1, in this segment, all available financial data on the project is sorted and formatted into a spreadsheet, Spreadsheet 11-2. This spreadsheet shows cash flows for years 1, 3, 5, 7 and 10. Comprehensive financial analyses over the entire 10-year period are shown in Appendix B. The $50,000 write-off of equipment being replaced is captured, in entirety, as negative cash flow in Year 1. Since the project is funded in entirety from the institution's capital budget, there are no annual payments in this scenario. Instead the entire investment of $2,100,000 is entered as a negative cash flow, accrued in Year 1. For simplicity, once again, all maintenance—including software and firmware updates—is assumed to be included in the overall cost of the project. The comprehensive or full version of Spreadsheet 11-2, as included in Appendix B, shows all amounts in full. The annual stipulated savings of $350,000 per year for this project are shown as positive cash flows for each year.

The present values for each of the 10 years are added together to yield the NPV, net present value, of the project. The NPV and other calculated financial indicators for the capital budget approach to the project are shown in the second part of Spreadsheet 11-2.

Spreadsheet 11-2. Financial Analysis of Investment on Energy/Utility Project Without EPD ESCO Approach—Project Financed Through Capital Budget—Scenario I-B.

Description	Year 1	Year 3	Year 5	Year 7	Year 10
	In $1,000s				
Costs					
Write-off of Existing Investment,	($50)				
Capital Investment	($2,100)				
Savings					
Annual Savings	$350	$350	$350	$350	$350
Net Annual Cash Flows	($1,800)	$350	$350	$350	$350
Present Values of Annual Cash Flows	($1,666.7)	$277.8	$238.2	$204.2	$162.1

Financial Analyses Results	
NPV	$357,788
IRR	6%
ROI	17%
Payback	6 Years

The IRR, Internal Rate of Return, for scenario I-B:
IRR = ($135,000)/($2,100,000) = 6%

The ROI, or simple Return on Investment, for I-B:
ROI = ($350,000)/($2,100,000) = 17%

The Payback Period for I-B:
Payback Period = ($2,100,000)/($350,000) = 6 years

Examination of the conventional financial indicators for the non-ESCO capital budget funding of the project reveals that this approach is financially stronger than the loan approach. The NPV for the project increases by 76% higher than the loan alternative. While the IRR, at 6%, is still low by most corporate standards, the ROI of 17% is competitive. The payback period of 6 years is still somewhat long as compared with productivity improvement projects in the corporate world. However, in the assessment of scenario I-B's projected financial performance, the following facts need to be borne in mind:

a) The overall energy/utility project consists of a combination of high return "low hanging" fruit and the low return "hard to reach" fruit. In other words, the results reflect financial performance offset of lower return measures by higher return measures. As discussed earlier in this chapter, this is a prudent financial strategy for "floating" the longer payback measures like chillers and chilled water system modifications by coupling them with higher financial performance, short payback, measures like lighting improvements.

b) Energy/utility projects show a far superior performance in geographic regions where the cost of energy and utilities is high.

c) Since this scenario, similar to I-A, does not involve an ESCO, the overall project cost is higher than what would be expected through a building or controls equipment type ESCO. This is evidenced by the lower overall investment cost in the tax-exempt lease-purchase ESCO scenario represented in Case Study 11-3.

CASE STUDY 11-3

II. This case study, and associated financial analysis, are premised on the ESCO approach. Since ESCOs bear most of the risk in most EPC ESCO projects, in this case study we will examine the financial performance of the energy/utility project from ESCO vantage point. We will do so by comparing the following three approaches to an energy/utility efficiency project being implemented through a tax-exempt lease-purchase EPC ESCO alternative:

A. EPC ESCO approach consisting of a combination of short-term payback projects and long-term payback projects

B. EPC ESCO approach consisting, solely, of high return, short payback, measures.

C. EPC ESCO approach consisting, exclusively, of low return, long payback period, measures.

II-A. Financial Analysis of Investment by ESCO on Tax-Exempt Lease-Purchase Energy/Utility Project—Low and High Return Measures Combined

Assumed Annual Discount (Interest) Rate: 8%

Assumed Tax-Exempt Lease-Purchase Lending/Borrowing Rate: 3%

Initial Cost of Project: $1,680,000

Project or Equipment Life = 10 Years

As with Case Study 11-2, all available financial data on the project is sorted and formatted into a spreadsheet, Spreadsheet 11-3, for segment II-A. This spreadsheet shows cash flows for years 1, 3, 5, 7 and 10. Comprehensive financial analysis over the entire 10-year period is shown in Appendix B. The $50,000 write-off of equipment being replaced is captured, in entirety, as negative cash flow in Year 1. Since the project in this scenario is to be implemented, turnkey, through a building and control equipment manufacturing ESCO, the project cost is assumed to be reduced by 20%; down from $2,100,000 to $1,680,000. The approximate 20% reduction could be explained on the following basis:

a) Overall project cost, including equipment, software, engineering, construction, installation and maintenance, would tend to be lower when the project is implemented, on turnkey basis, by a building and control equipment manufacturing ESCO.

b) Economies of scale, low or no mark-up on equipment and services, since most of the equipment may be manufactured or distributed by the building and control equipment manufacturing ESCO.

c) Availability of internal engineering, maintenance and construction resources and expertise.

As shown in Spreadsheet 11-3, there are two different interest rates involved in this scenario. The first rate is the prevailing discount rate of 8%, stated above. This rate is to be utilized mainly for present value and NPV calculations. This rate is the, assumed—current—"market" rate, at which most commercial borrowing and lending is conducted. The second rate employed in the financial analysis of this case study is the assumed 3% financing rate. This rate is, substantially, lower than market rate because of the following reasons:

a) Since this project is a tax-exempt lease-purchase type ESCO project, the lender would be exempt from paying federal income taxes on the interest earned.

b) Tax-exempt lease-purchase type ESCO projects are regarded as, relatively, low risk and secured to a certain extent by tangible assets; another reason for the lender to lower the interest rate.

c) Relative higher probability of success on tax-exempt lease-purchase type ESCO projects, based on historical data.

d) Greater confidence in the projected financial performance due to the use of stipulated and proven savings/revenues.

For simplification purposes, once again, all maintenance—including software and firmware updates—is assumed to be included in the overall cost of the project. The comprehensive or full version of Spreadsheet 11-3, as depicted in Appendix B, shows all amounts, in full. The annual savings of $350,000 per year, projected and stipulated for this project, are shown as positive cash flows for each year. Also, to simplify the case study, it is assumed that the contract sets the savings distribution threshold at $350,000. This implies that, in any given year, the ESCO is not obligated to share the savings from the project with the client institution unless the level of savings exceeds $350,000. By the same token, this also implies that if, in a given year, the savings or revenue fall short of the $350,000 mark, the financial performance of the project, for that particular year, would be lower than the figures stated in Spreadsheet 11-3.

The NPV and other calculated financial indicators for the tax-exempt lease-purchase approach to the project are shown in the second part of Spreadsheet 11-3.

Spreadsheet 11-3. Financial Analysis of Investment by ESCO on tax-exempt lease-purchase Energy/Utility Project—Low and High Return Measures Combined. Scenario II-A.

Description	Year 1	Year 3	Year 5	Year 7	Year 10
	In $1,000s				
Costs					
Write-off of Existing Investment,	($50)				
Annual Payments on tax-exempt lease-purchase ESCO Loan	($196.9)	($196.9)	($196.9)	($196.9)	($196.9)
Savings					
Annual Savings	$350	$350	$350	$350	$350
Net Annual Cash Flows	$103.10	$153.10	$153.10	$153.10	$153.10
Present Values of Annual Cash Flows	$95.467	$121.54	$104.2	$89.335	$70.917

Financial Analyses Results	
NPV	$981,044
IRR	9%
ROI	9%
Payback	10.97 Years

The IRR, Internal Rate of Return, for II-A:
IRR = ($148,104)/($1,680,000) = 9%

The ROI, or simple Return on Investment, for II-A:
ROI = ($153,104)/($1,680,000) = 9%

The Payback Period for II-A:
Payback Period = ($1,680,000)/($148,104) = 10.97 years

Examination of the financial indicators for the tax-exempt lease-purchase ESCO approach shows that this approach is financially strong, for an energy/utility efficiency project. By corporate standards, the IRR and ROI ratios are strong enough to meet most thresholds or standards in the energy arena. The $981,044 NPV for this scenario is remarkably strong. From ESCO's perspective, this substantial NPV value implies that this energy/utility productivity enhancement ESCO project will not only pay for itself within the 10-year lifespan of the assets but will add, approximately, $981,044 to the ESCO's bottom line.

Other observations that assist in examining the results of scenario II-A analyses in proper and more comprehensive perspective are as follows:

a) The overall energy/utility project consists of a combination of high return and the low return measures.

b) Energy/utility projects would show a far superior performance in geographic regions where the cost of energy and utilities is high.

II-B. Financial Analysis Of Investment By ESCO On Tax-Exempt Lease-Purchase Energy/Utility Project—High Return Measures Only

Assumed Annual Discount (Interest) Rate: 8%

Assumed Tax-Exempt Lease-Purchase Lending/Borrowing Rate: 3%

Initial Cost of Project: $800,000

Project or Equipment Life = 10 Years

Segments II-B and II-C are designed to illustrate the importance of combining short payback and long payback energy/utility efficiency measures in the effort to sustain the lower return capital intensive measures. Case Study 11-3, as a whole, shows that the mission of achieving overall energy/utility efficiency in facilities depends heavily on bundling the low and high return measures into a single project. And, as mentioned in the discussion for Scenario II-A, Scenarios II-B and II-C represent the ESCO vantage point. The financial performance results in Spreadsheets 11-4 and 11-5 indicate the financial advantage or disadvantage for the ESCO, and not the customer or institution.

In Segment II-B, we will include only the higher return measures from Table11-1. Measures such as lighting system upgrade and application VFD application are the only measures included in this scenario. The modified cost estimate is shown in Table 11-2. Since the project, with the scope limited to the high return measures, would still be implemented, turnkey, through a building and control equipment manufacturing ESCO, the 20% reduction, as explained in scenario II-A, would still apply. The total investment with high return measures, exclusively, as noted in Table 11-2 is $800,000. The savings pertaining to the high return measure scenario are $225,000.

Table 11-2. Costs and Savings Associated with High Return Measures

No.	Description	Measure Cost	Annual Savings
1	Lighting System Upgrade	$900,000	$200,000
2	Installation of VFDs	$100,000	$25,000
	Subtotals	$1,000,000	$225,000
	20% ESCO Discount/ Cost Advantage	$200,000	
	Net Investment	$800,000	

All pertinent financial data for segment II-B is shown in Spreadsheet 11-4. Comprehensive financial analyses over the entire 10-year period are included in Appendix B, for reference.

Spreadsheet 11-4. Financial Analysis Of Investment By ESCO On Tax-Exempt Lease-Purchase Energy/Utility Project—High Return Measures Only. Scenario II-B.

Description	Year 1	Year 3	Year 5	Year 7	Year 10
	In $1,000s				
Costs					
Write-off of Existing Investment,	($50)				
Annual Payments on Tax-Exempt Lease-Purchase ESCO Loan	($93.76)	($93.76)	($93.76)	($93.76)	($93.76)
Savings					
Annual Savings	$225	$225	$225	$225	$225
Net Annual Cash Flows	$81.240	$131.24	$131.24	$131.24	$131.24
Present Values of Annual Cash Flows	$75.222	$104.183	$89.335	$76.57	$60.790

Financial Analyses Results	
NPV	$834,335
IRR	16%
ROI	16%
Payback	6 Years

It is assumed that the contract sets the savings distribution threshold for scenario II-B at $225,000.

The NPV and other calculated financial indicators for scenario II-B are shown in the second part of Spreadsheet 11-4.

The IRR, Internal Rate of Return, for II-B:
IRR = ($126,240)/($800,000) = 16%

The ROI, or simple Return on Investment, for II-B:
ROI = ($131,240)/($800,000) = 16%

The Payback Period for II-B:

Payback Period = ($800,000)/($126,240) = 6.1 years

Examination of the financial indicators for exclusive high return measure tax-exempt lease-purchase ESCO approach shows that this approach is financially the strongest. As obvious, this scenario outperforms the combined high and low return measure approach analyzed under scenario II-A. All financial ratios for the II-B scenario are higher than their counterparts in scenario II-A. By corporate standards, the IRR and ROI ratios for II-B are strong enough to meet most thresholds of approval in the energy market. The $834,335 NPV for this scenario is strong. From ESCO's perspective, this substantial NPV value implies that this energy/utility productivity enhancement ESCO project, consisting solely of high return measures, will not only pay for itself within the 10-year lifespan of the assets but will add, approximately, $834,335 to the ESCO's bottom line.

While the IRR and ROI, at 16%, are competitive even by corporate standards, the payback period of 6 years is still somewhat long as compared with productivity improvement projects in the corporate world. Note that this scenario would show a far superior performance in geographic regions where the cost of energy and utilities is high.

II-C. Financial Analysis Of Investment By ESCO On Tax-Exempt Lease-Purchase Energy/Utility Project—Low Return Measures Only

Assumed Annual Discount (Interest) Rate: 8%

Assumed Tax-Exempt Lease-Purchase Lending/Borrowing Rate: 3%

Initial Cost of Project: $880,000

Project or Equipment Life = 10 Years

In Segment II-C, we will include only the lower return, long payback period, measures from Table11-1. Automated HVAC, EMS, cooling tower, chiller and chilled water system modification are the only measures included in this scenario. The modified cost estimate is shown in Table 11-3. Since the project, with the scope limited to the low return measures, would still be implemented, turnkey, through a building and control equipment manufacturing ESCO, the 20% reduction, as explained in scenario II-A and II-B, would still apply. The total investment with low return measures, exclusively, as noted in Table 11-3 is $880,000. The savings pertaining to the low return measure scenario are $125,000.

Table 11-3. Costs and Savings Associated with Low Return Measures

No.	Description	Measure Cost	Annual Savings
1	Automated HVAC Control & EMS Systems	$400,000	$40,000
2	Replacement of Cooling Tower	$100,000	$10,000
3	Replacement of a chiller	$300,000	$40,000
4	Expansion of Existing Chilled Water System.	$300,000	$35,000
	Subtotals	$1,100,000	**$125,000**
	20% ESCO Discount/ Cost Advantage	$220,000	
	Net Investment	**$880,000**	

All pertinent financial data for segment II-C is shown in Spreadsheet 11-5. Comprehensive financial analyses over the entire 10-year period, as before, are included in Appendix B.

Spreadsheet 11-5. Financial Analysis of Investment by ESCO on Tax-Exempt Lease-Purchase Energy/Utility Project—Low Return Measures Only. Scenario II-C.

Description	Year 1	Year 3	Year 5	Year 7	Year 10
	In $1,000s				
Costs					
Write-off of Existing Investment,	($50)				
Annual Payments on Tax-Exempt Lease-Purchase ESCO Loan	($103.1)	($103.1)	($103.1)	($103.1)	($103.1)
Savings					
Annual Savings	$125	$125	$125	$125	$125
Net Annual Cash Flows	($28.14)	$21.86	$21.86	$21.86	$21.86
Present Values of Annual Cash Flows	($26.05)	$17.36	$14.88	$12.757	$10.127

Financial Analyses Results	
NPV	$100,413
IRR	2%
ROI	2%
Payback	40.25 Years

It is assumed that, by contract, the savings distribution threshold for scenario II-C is set at $125,000.

The NPV and other calculated financial indicators for scenario II-C are shown in the second part of Spreadsheet 11-5.

The IRR, internal rate of return, for II-C:
IRR = ($16,864)/($880,000) = 2%

The ROI, or simple return on investment, for II-C:
ROI = ($21,864)/($880,000) = 2%

The payback period for II-C:
Payback period = ($880,000)/($16,864) = 40.25 years

Examination of the financial indicators for the low return measure tax-exempt lease-purchase ESCO approach shows that this approach is financially the weakest. The financial performance of this scenario, from ESCO's vantage point, is substantially weaker than the combined high and low return measure approach analyzed under scenario II-A. All financial ratios for the II-C scenario are lower than their counterparts in scenario II-A and II-B. By corporate standards, the IRR and ROI ratios for II-B are not strong enough to meet most thresholds of approval in the energy market. The $100,413 NPV for this scenario, while still positive, is the lowest. The payback period of 40.25 years is unacceptably long.

In conclusion, Case Study 11-3 financial analyses demonstrate that bundling of low return and high return measures constitutes a viable strategy in the accomplishment of overall energy/utility productivity improvement mission.

SELF ASSESSMENT PROBLEMS AND QUESTIONS—CHAPTER 11

Problem 1.
Solve Case Study problem 11-1 (a) and (b) for a demand reduction of 9%.

Problem 2.
Evaluate the NPV and the financial ratios in Case Study 11-2, segment I-B, for annual cost savings of $400,000.

Problem 3.
Evaluate the NPV and the financial ratios in Case Study 11-3, segment II-A, for annual cost savings of $400,000.

Appendix A

Solutions and Answers

To End-of-Chapter Self Assessment Problems and Questions

CHAPTER 1

1. B. False
2. A. 2%.
3. B. 16%
4. B. False.

The term annuity applies to payments involving periods other than annual periods, as well.

5. **Solution:**

Bond Price = ($F_{\text{Par Value of Bond}}$) • (*P*/*F*, i%, n) + ($A_{\text{Interest Payment}}$) • (P/A, i%, n)

Bond Price = ($5,000). (*P*/*F*, 10%, 10) + (400) • (P/A, 10%, 10)

Bond Price = ($5,000). (0.3855) + (400). (6.1446)

Bond Price = $1,927.50 + $3,686.76 = $5,614.26

Answer: D or $5,614

6. **Solution**:

(a) Applying Eq. 1-1:

$$\text{Discount Yield (\%)} = \frac{(\$10{,}000 - \$9{,}200)}{\$10{,}000} \times \frac{360}{90} \times 100\%$$

Discount Yield (%) = 32%

Answer: B.32%

(b) Applying Eq. 1-2:

Yield to Maturity (%) =

$$\frac{(\$10,000 - \$9,200)}{\$9,200} \times \frac{365}{90} \times 100\%$$

Yield to Maturity (%) = 35 %
Answer: A. 35%

7. B. Cost Center

8. D. Both (A) and (C)

9. (a) **Solution**:
Sales: $150,000
Cost of Goods Sold:
Inventory at the beginning of the month: $10,000
Purchases: $60,000
Direct Labor: <u>$40,000</u>
Subtotal: $110,000
Less: Inventory at the end of the billing month: $10,000
Net Cost of Goods Sold: $110,000 – $10,000 = $100,000
Answer: A. $100,000

(b **Solution**:
Gross Profit on Sales = Sales – COGS
= $150,000 – $100,000
= $50,000
Answer: A. $50,000

10. C. Organization II

CHAPTER 2

1. A. True

2. **Solution**:
According to Equation 3a,
Break Even Volume = $v_{BE} = c_f/(p - c_v)$
Revised data from Smith's Automotive, in accordance with the statement of this problem, are as follows:

c_f = \$30,000
c_v = \$3 per pair
p = \$20 per pair
v_{BE} = cf/(p – cv)
v_{BE} = \$30,000/(\$20 – \$3)
= 1,765 pairs
Answer: C, 1,765 pairs.

3. D. Building Lease Payment

4. B. False

CHAPTER 3

Problem/Question—1
Solution:

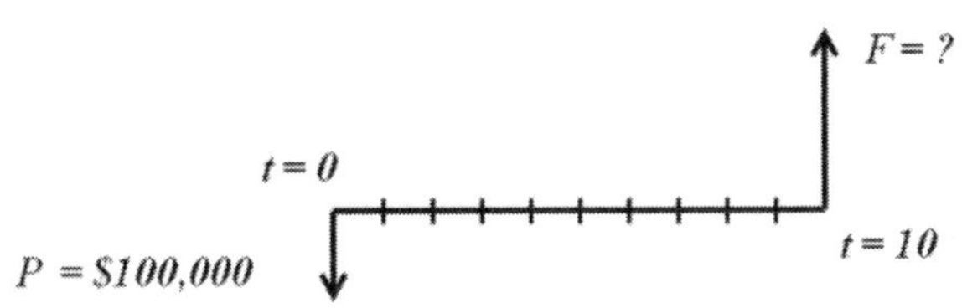

F = P (*F/P*, 15%, 10)
= \$100,000 (4.046)
= \$404,600
Answer: C

Problem/Question—2
Solution:

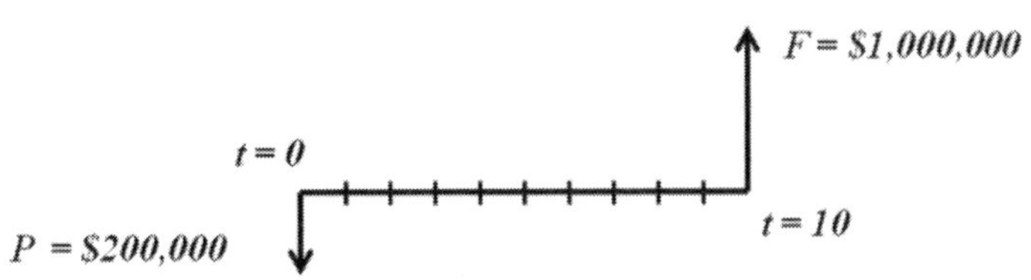

F = P (*F/P*, ?%, 10), or
$F = P(1 + i)^n$

$(F/P)^{1/n} - 1 = i$
$i = (\$1{,}000{,}000/\$200{,}000)^{1/10} - 1$
$i = 0.17$ or 17%
Answer: D

Problem/Question—3
Solution:

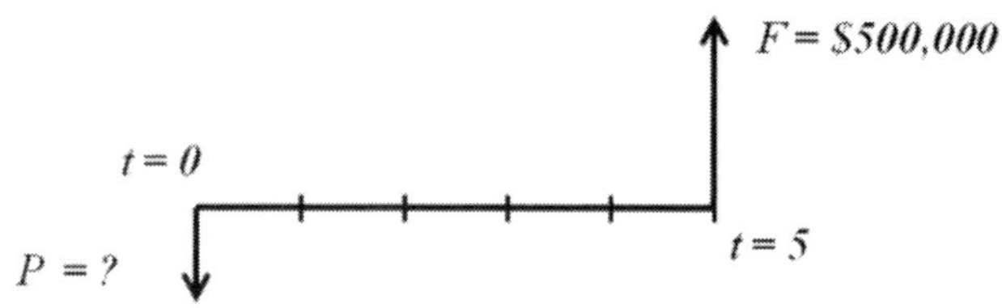

$P = F\,(P/F, 20\%, 5)$
$= \$500{,}000\ (0.4019)$
$= \$200{,}950$
Answer: A

Problem/Question—4
Solution:

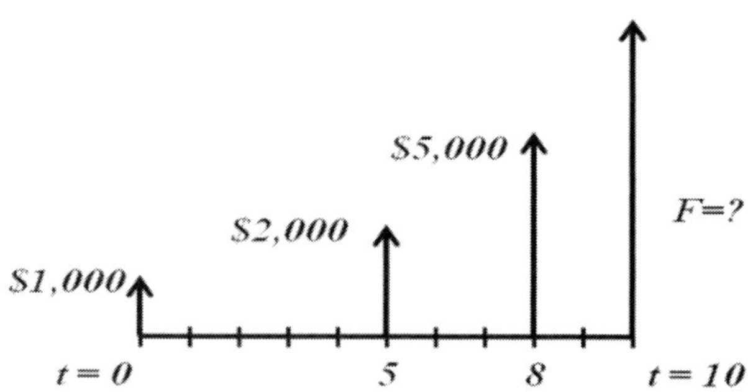

$F_{Total} = P_0 \,.\, (F/P, 10\%, 10) + P_5 \,.\, (F/P, 10\%, 5) + P8 \,.\, (F/P, 10\%, 2)$

$F_{Total} = (\$1{,}000) \,.\, (F/P, 10\%, 10) + (\$2{,}000) \,.\, (F/P, 10\%, 5) + (\$5{,}000) \,.\, (F/P, 10\%, 2)$

$F_{Total} = \$1{,}000 \,.\, (2.594) + (\$2{,}000) \,.\, (1.611) + (\$5{,}000) \,.\, (1.210)$; using financial factors from Appendix A

F_{Total} = \$11,866

Answer: B

Problem/Question—5
Solution:

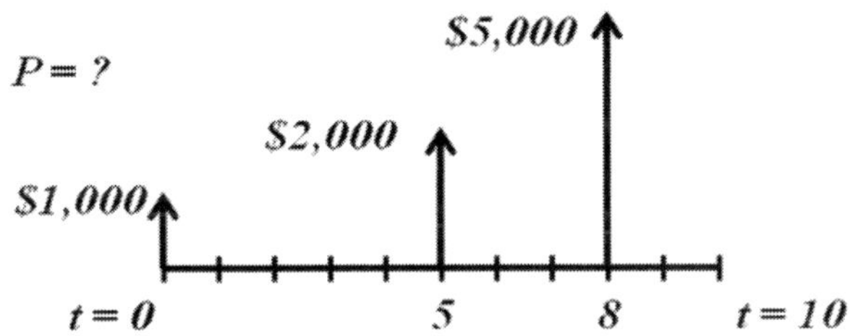

$P_{Total} = F_0 \cdot (P/F, 10\%, 0) + F_5 \cdot (P/F, 10\%, 5) + F_8 \cdot (P/F, 10\%, 8)$

$P_{Total} = (\$1{,}000) \cdot (P/F, 10\%, 0) + (\$2{,}000) \cdot (P/F, 10\%, 5) + (\$5{,}000) \cdot (P/F, 10\%, 8)$

$P_{Total} = \$1{,}000 \cdot (1) + (\$2{,}000) \cdot (0.6209) + (\$5{,}000) \cdot (0.4665)$; using financial factors from Appendix A

P_{Total} = \$4,574

Answer: D

Problem/Question—6

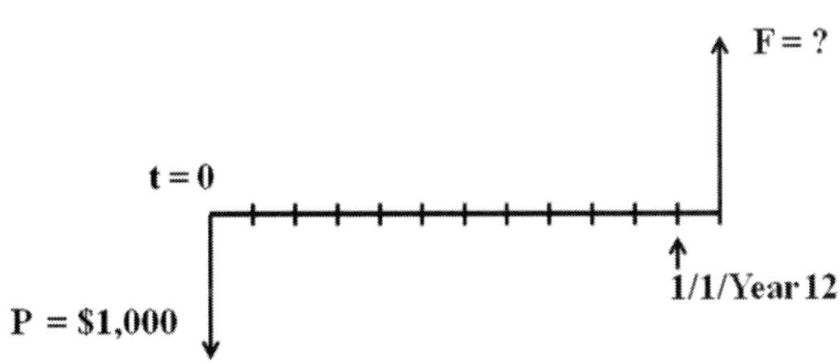

Solution:

Since only 11 years worth of interest is accrued by January 1, Year 12, n = 11. Therefore,

$$F = P\,(F/P\,, 9\%, 11)$$
$$= \$1{,}000\ (2.5804)$$
$$= \$2{,}580$$

Answer: C

Problem/Question—7
Solution:

	Project "X"	Project "Y"
Initial Cost	\$22,000	\$27,000
Annualized Cost of Operation	\$3,100	\$2,600

NPV for Project X or Y = PV of Initial Cost + PV of Annualized Cost of Operation

NPV(X) = –\$22,000 – A (P/A, 25%, 25)

NPV(X) = –\$22,000 – \$3,100 (3.9849) = – \$34,353

And, NPV(Y) = –\$27,000 – A (P/A, 25%, 25)

NPV(Y) = –\$27,000 – \$2,600 (3.9849) = – \$37,361

Since the NPV for Project or Alt. X is less negative (lower cost), select "X."

Answer: A—Project X

Problem/Question—8
Solution:

Option A:

F = A. (F/A, 1%, 24) = (\$100) . (26.9735) = \$2,697.35

Option B:

F = (Annuity). (F/A, 12%, 2) = (\$1,200) . (2.1200) = \$2,544

Answer: Option A

CHAPTER 4

1. A. True
2. B. False
3. A. True
4. A. 8-K report
5. A. True
6. B. False

CHAPTER 5

1. A. True
2. D. Annual Income Statement
3. C. 303%; Percent change in Net Income, from 2003 to 2005 = $46.83 MM/$15.48 MM x 100%= 303%
4. A. $1,634.15 MM
5. A. $164,700,000
6. A. 37.97%. The income tax bracket for ABC Corp., based on 2004 data, is = $34.37MM/$90.53 MM x 100% = 37.97%
7. A. True. See the Liabilities section of the Balance Sheet.
8. B. $9.22. Price per unit = $922.12 MM/100MM = $9.22
9. A. Duke is approximately 50% larger.
   Size Ratio = (Duke's 2010 Total Op. Rev.)/(ABC 2005 Net Sales) = $3,594 MM/$2,359.45 MM = 152%
10. B. ABC shows greater growth based on net sales or revenue.
    Duke: Growth Ratio = $3,594 MM/$3,312 MM = 109%
    ABC: Growth Ratio = $2,359.45 MM/$1,417.19MM = 166%
11. D. Both A and C
    Duke: Growth Ratio = $445 MM/$349 MM = 128%
    ABC: Growth Ratio = ($46.83 MM – $55.78 MM)/$46.88 MM = –19%
12. B. $17,787 MM. See asset portion of Duke's Balance Sheet.

CHAPTER 6

1. A. Yes
2. B. False

3. B. False
4. D. $22,219.

Solution:

Annual energy cost for operating the new, energy efficient, equipment in Case Study 6-1 = $32,675 – $10,456 = $22,219

5. D. 0.59

Solution:

Plant-turnover Ratio = Sales Revenue/Investment
= $2,359.45 MM/$4,022.09MM
= 0.59

6. D. 3.06

Solution:

Inventory Turnover Ratio = Cost of Goods Sold/Average Inventory
= $811.11 MM/$264.73 MM
= 3.06

7. C. $1,637,000,000

Solution:

Working Capital = Current Assets – Current Liabilities
Working Capital = $5,491 MM – $3,854 MM = $1,637,000,000

8. B. 68.3%

Solution:

D/E = Debt (or Liabilities)/Equity
= $14,953 MM/$21,886 MM
= 0.683 or 68.3%

CHAPTER 7

1. B. Straight Line Method
2. D. $400,000
3. A. $989 MM

Solution:

Operating Income QE March 31, 2010
= $3,594 MM+2MM-($2,835 MM-$456 MM)-($456 MM) . (0.5)
= $989 MM

4. D. Results A & B
5. A. True
6. B. False
7. B. False

CHAPTER 8

1. **Answer**: A. 3.06
   ITR = Cost of Goods Sold ($)
   Average Inventory ($)
   = (811.11/264.73) = 3.06
2. A. True
3. B. False
4. D. All of the above
5. A. True
6. B. False
7. A. True
8. D. 447 units
9. **Solution**:
   (a) **Answer**: C. $6,000
   Annual Carrying Cost = Cc .Q/2 = ($30) • (400/2) = $6,000

   (b) **Answer**: D. $1,500
   Annual Ordering Cost = Co . D/Q = ($60) • (10,000/400) = $1,500

   (c) **Answer**: A. 12 days
   Time between orders, or Order Cycle, in days = Q opt/D
   = (400/10,000 per year) x 300 days/year
   = 12 days

10. **Answer**: D. 205 trucks

Solution:

Amount of coal needed per year = 1,500,000 tons
Number of trucks or railcars needed per year = 75,000
Number of trucks or railcars needed per day = 75,000/365 = 205

CHAPTER 9

1. B. False
2. A. True
3. B. False
4. B. $505,000

Solution:

Demand Cost for the month = $14.25/kW x 5000 kW + $13.25/kW x 5000 kW + $12.25/kW x 30,000 kW = $505,000

5. C. $1,064,187

Solution:

Baseline Billing Determinants	**Data**	**Data**	**Calculated**
Billing Demand		26,000	26,00 kW
On-Peak Billing Demand		26,000	26,00 kW
On-Peak Energy Usage		4,334,573	4,334,573
Off-Peak Energy Usage		14,500,000	14,500,000
Baseline Charges			
Basic Facilities Charge		$36.0700	$36.07
Extra Facilities Charge			$13,000.00
On-Peak Billing Demand Charge			
For the First 2,000 kW	2,000	$14.0080	$28,016.00
For the Next 3,000 kW	3,000	$12.8319	$38,495.70
For All Over 5,000 kW	21,000	$11.6449	$244,542.90
Economy Demand Charge	—	**$1.1136**	**$0.00**
On-Peak Energy Usage	4,334,573	$0.0570	$247,183.36
Off-Peak Energy Usage	14,500,000	$0.0340	$492,913.00
Total Baseline Charge		**$0.0565**	**$1,064,187.03**

HP Billing Determinants			
Total Actual Demand For The Facility	40,000		
HP Charges			
Incremental Demand Charge	14,000	$—	$0.00
New Load x Avg. Hourly Price	8,805,800	$—	$0.00
Total HP Charge			**$0.00**
Total Bill for the Month:			**$1,064,187.03**

6 A. $672,000

Solution:

Type of Charge	**Ccf**	**Charge per Ccf**	**Total**	
Total Gas Bill			$671.66	
Monthly Service Charge			$10.20	$10.20
Volume Charge for the first 500 Ccf	500	$0.43	$216.50	$216.50
Volume Charge for additional usage over 500 Ccf	50	$0.42	$42.30	$21.15
Purchased Gas Cost Charge	550	$0.74	$441.00	$404.25
Billing Period: 2/24/2010-3/22/2010				$652.10
Sales Tax: $3.55 (at %)			$19.56	
Total Amount Due (Bill for the month): $122.04				**$671.66**

CHAPTER 10

1. B. False
2. A. True
3. A. Direct Cost

4. C. Uncontrollable cost.
5. A. Negative Variance
6. D. Both A and B
7. B. False
8. B. False
9. A. True
10. D. Energy Cost
11. B. False
12. A. Replace Scenario $6,000
13. A. True

CHAPTER 11

Problem/Question—1

Solution:

Part (a): Based on the demand elasticity model depicted in Figure 11-1, estimate annual electrical energy cost reduction that could, potentially, be realized though this joint collaboration between the industrial consumer and the utility if a 9% peak demand reduction is achieved through Demand Response Measures.

Current On-peak Energy Usage per year, in kWh
= 87,600,000 kWh

Current On-peak Energy Cost per year
= $11,388,000

Projected On-peak Demand, in MW, after implementation of the Demand Response Project
= Q_2

= (30 MW) x (1 – 0.09)
= 27.3 MW

Projected On-peak Energy Usage per year, in kWh
= (27.3 MW) x (1,000,000 W/MW) x (1 kW/1000 W) x (8 hours/

day) x (365 days/year)
= 79,716,000 kWh

Projected % Reduction in Energy Cost Rate due to 9% Demand Reduction, through Demand Response Measures, based on California Electricity Crisis model of 2000-2001

= ΔP (%)
= (50% Price Reduction/5% Demand Reduction) x (9% Demand Reduction in this Case Study)
= 90%

Projected On-peak Energy Cost Rate
= P_2
= (1 – 0.9) x ($0.13/kWh)
= $0.013/kWh

Projected On-peak Energy Cost per year
= (79,716,000 kWh) x ($0.013/kWh)
= $1,036,308

Therefore, the projected annual energy cost reduction due to 9% demand reduction, through demand response measures

= $11,388,000 – 1,036,308
= $10,351,692

Part (b): What would the annual cost reduction be, in percent (%), due to the 9% on-peak demand reduction?

Annual Cost Reduction, in %
= ($11,388,000 – 1,036,308)/($11,388,000) x 100%
= 90.9%

Problem/Question—2:
Solution:

NPV (2)	$693,292
IRR	9%
ROI:	19%
Payback Period, in YEARS:	5.25

Problem/Question—3

Solution:

NPV (2)	$1,316,548
IRR	12%
ROI:	12%
Payback Period, in YEARS:	8.27

Appendix B

Comprehensive Microsoft Excel ® Spreadsheets

For Referenced Case Studies and Problems

Spreadsheet 6.1 - Case Study 6.1

Financial Analysis of Capital Budget Financed Equipment Replacement										
1. Annual Utilities Cost, @ $0.05/kWh; 24/7 Operation for Equipment to be Replaced:		$ (32,675)		Assumed Annual Discount (Interest) Rate:		10%				
Initial Cost of Project: $ 200,000		Proj./Eq. Life	10	Years						
Description:	Year	1	2	3	4	5	6	7	9	10
COSTS:										
1. Write-off of Existing Investment, at Book Value		$ (10,000)								
2. Total Initial Cost of the Project, Funded Through Capital Budget		$ (200,000)								
SAVINGS:										
1. Utilities / Energy: Eff. Improvement.	32%	$ 10,456	$ 10,456	$ 10,456	$ 10,456	$ 10,456	$ 10,456	$ 10,456	$ 10,456	$ 10,456
2. Productivity Improvement:	35%	$ 11,436	$ 11,436	$ 11,436	$ 11,436	$ 11,436	$ 11,436	$ 11,436	$ 11,436	$ 11,436
3. Est. Annual Maint. Cost Reduction:		$ 30,000	$ 30,000	$ 30,000	$ 30,000	$ 30,000	$ 20,000	$ 20,000	$ 20,000	$ 20,000
4. Reduced Safety Incident Rate, Shoulder, Back and Arm Injuries (Based on Historical Data)		$ 10,000	$ 10,000	$ 10,000	$ 10,000	$ 10,000	$ 10,000	$ 10,000	$ 10,000	$ 10,000
Net Annual Cash Flows:		$ (148,108)	$ 61,892	$ 61,892	$ 61,892	$ 61,892	$ 51,892	$ 51,892	$ 51,892	$ 51,892

Spreadsheet 6.1 - Case Study 6.1, Contd.										
Description:	Year	1	2	3	4	5	6	7	9	10
Net Annual Cash Flows:		$ (148,108)	$ 61,892	$ 61,892	$ 61,892	$ 61,892	$ 51,892	$ 51,892	$ 51,892	$ 51,892
PV's, Present Values:		$ (134,644)	$51,151	$46,500	$42,273	$38,430	$29,292	$26,629	$22,007	$20,007
NPV (1)		$161,188	Sum of ALL Present Values from the ten Year Life of the Equipment or Project							
IRR		17%	Average of ALL Cash Flows, or Net Savings Generated By the Project, During its Ten Year Life Expectancy Divided by the Total Investment							
ROI:		31%	Net Inc./Avg.Total Assets:							
Payback Period, in YEARS:		3.62	Total Investment Divided By the Average of Cash Flows							

* Note:

Savings Based on Reduced Safety Incident Rate Include The Following Cost Assumptions:

1) Approx. Cost for Back Surgery:		$ 25,000		
2) Approx. Cost for Carpel Tunnel Surgery:		$ 15,000		
3) Approx. Cost for Shoulder Surgery:		$ 25,000		
4) Approx. Cost Disability Settlements:		$ 120,000		
	Total	$ 185,000		
However, for this calculation, the annual cost assumed for this category:			$ 10,000	36250

Case Study 10.2 - Life Cycle Cost Comparison Based on Time Value of Money

Discount Rate: 10%

Scenario (1): Rewind Twice, then Replace at 3rd Failure:

	Year #1	Year #2	Year #3	Year #4	Year #5	Year #6	Year #7	Year #8	Year #9	Year #10
Actual Energy Cost:	$ (34,036)	$ (34,036)	$ (34,036)	$ (34,036)	$ (34,036)	$ (35,828)	$ (35,828)	$ (35,828)	$ (37,713)	$ (37,713)
Purchase & Rewind Costs:	$ (3,400)					$ (1,600)		$ (1,600)		
Lost Prod & Maint. Cost:						$ (3,000)		$ (3,000)		$ (3,000)
Cash Flows:	$ (37,436)	$ (34,036)	$ (34,036)	$ (34,036)	$ (34,036)	$ (40,428)	$ (35,828)	$ (40,428)	$ (37,713)	$ (40,713)
NPV of All Costs:	($223,871)									

Scenario (2): Replace Upon 1st Failure:

	Year #1	Year #2	Year #3	Year #4	Year #5	Year #6	Year #7	Year #8	Year #9	Year #10
Actual Energy Cost:	$ (34,036)	$ (34,036)	$ (34,036)	$ (34,036)	$ (34,036)	$ (34,036)	$ (34,036)	$ (34,036)	$ (34,036)	$ (34,036)
Purchase & Rewind Costs:	$ (3,400)					$ (3,400)		$ -		
Lost Prod & Maint. Cost:						$ (3,000)		$ -		$ (3,000)
Cash Flows:	$ (37,436)	$ (34,036)	$ (34,036)	$ (34,036)	$ (34,036)	$ (40,436)	$ (34,036)	$ (34,036)	$ (34,036)	$ (37,036)
NPV of All Costs:	($216,997)									

Cost Diff. Scenario (1) vs. Scenario (2): ($6,874)

Spread Sheet 11.1 Financial Analysis of Investment on Energy/Utility Project Without EPD ESCO Approach

I - A. Project Financing Through Cash Borrowed from Capital Budget; Charged to the Operations at the Discount Rate.

Assumed Annual Discount (Interest) Rate: 8%

Annual Payment Toward Amortization of Borrowed Capital = $2,100,000 (A/P,8%,10) = ($2,100,000). (0.1490) = $ 312,900

Initial Cost of Project: $ 2,100,000	Project/Equipmnet Life: 10 Years									
	Operating Years & Cash Flows									
Description:	1	2	3	4	5	6	7	8	9	10
COSTS:										
Write-off of Existing Investment, at Book Value	$ (50,000)									
Annual Payments on Capital Loan	$ (312,900)	$ (312,900)	$ (312,900)	$ (312,900)	$ (312,900)	$ (312,900)	$ (312,900)	$ (312,900)	$ (312,900)	$ (312,900)
SAVINGS:										
Annual Savings	$ 350,000	$ 350,000	$ 350,000	$ 350,000	$ 350,000	$ 350,000	$ 350,000	$ 350,000	$ 350,000	$ 350,000
Net Annual Cash Flows:	$ (12,900)	$ 37,100	$ 37,100	$ 37,100	$ 37,100	$ 37,100	$ 37,100	$ 37,100	$ 37,100	$ 37,100
PV's, Present Values:	($11,944)	$31,807	$29,451	$27,270	$25,250	$23,379	$21,647	$20,044	$18,559	$17,184
NPV (1)	$202,648	Sum of ALL Present Values from the ten Year Life of the Equipment or Project								
NPV (2)	$202,648	Excel Based Calculation								
IRR	2%	Average of ALL Cash Flows, or Net Savings Generated By the Project, During its Ten Year Life Expectancy Divided by the Total Investment								
ROI:	2%	Net Inc./Avg.Total Assets:								
Payback Period, in YEARS:	56.60	Total Investment Divided By the Average of Cash Flows								

Energy Project Cost and Savings Breakdown			
No.	Description	Measure Cost	Annual Savings
1	Lighting System Upgrade	$900,000	$200,000
2	Installation of VFD's	$100,000	$25,000
3	Automated HVAC Control & EMS Systems	$400,000	$40,000
4	Replacement of Cooling Tower	$100,000	$10,000
5	Replacement of a chiller	$300,000	$40,000
6	Expansion of Existing Chilled Water System.	$300,000	$35,000
	Totals	**$2,100,000**	**$350,000**

Spread Sheet 11.2 Financial Analysis of Investment on Energy/Utility Project Without EPC ESCO Approach

I - B. Project Financing Through Cash from Capital Budget

Assumed Annual Discount (Interest) Rate: 8%

Initial Cost of Project: $ 2,100,000	Project/Equipment Life: 10									
	Operating Years & Cash Flows									
Description:	1	2	3	4	5	6	7	8	9	10
COSTS:										
Write-off of Existing Investment, at Book Value	$ (50,000)									
Capital Investment	$ (2,100,000)									
SAVINGS:										
Annual Savings	$ 350,000	$ 350,000	$ 350,000	$ 350,000	$ 350,000	$ 350,000	$ 350,000	$ 350,000	$ 350,000	$ 350,000
Net Annual Cash Flows:	$ (1,800,000)	$ 350,000	$ 350,000	$ 350,000	$ 350,000	$ 350,000	$ 350,000	$ 350,000	$ 350,000	$ 350,000
PV's, Present Values:	($1,666,667)	$300,069	$277,841	$257,260	$238,204	$220,559	$204,222	$189,094	$175,087	$162,118
NPV (1)	$357,788	Sum of ALL Present Values from the ten year life of the equipment or project								
NPV (2)	$357,788	Excel Based Calculation								
IRR	6%	Average of ALL Cash Flows, or Net Savings Generated By the Project, During its Ten Year Life Expectancy Divided by the Total Investment								
ROI:	17%	Net Inc./Avg.Total Assets:								
Payback Period, in YEARS:	6.00	Total Investment Divided By the Average of Cash Flows								

Energy Project Cost and Savings Breakdown			
No.	Description	Measure Cost	Annual Savings
1	Lighting System Upgrade	$900,000	$200,000
2	Installation of VFD's	$100,000	$25,000
3	Automated HVAC Control & EMS Systems	$400,000	$40,000
4	Replacement of Cooling Tower	$100,000	$10,000
5	Replacement of a chiller	$300,000	$40,000
6	Expansion of Existing Chilled Water System.	$300,000	$35,000
	Totals	$2,100,000	$350,000

Spread Sheet 11.3. Financial Analysis Of Investment By ESCO On Tax-Exempt Lease-Purchase Energy/Utility Project

- Low And High Return Measures Combined. Scenario II-A.

II - A. Project Financing Through Cash Borrowed from Commercial Financial Institution at a Lower Tax-Exempt Rate

Asssumed Annual Interest Rate for Tax-Exempt Lease-Purchase Financing of the Project: 3%

Assumed Annual Discount (Interest) Rate: 8%

Annual Payment Toward Amortization of Borrowed Capital = $1,680,000 (A/P, 3%,10) = ($1,680,000). (0.1172) = $ 196,896.00

Initial Cost of Project: $ 1,680,000	Project/Equipmnet Life: 10									
	Operating Years & Cash Flows									
Description:	1	2	3	4	5	6	7	8	9	10
COSTS:										
Write-off of Existing Investment, at Book Value	$ (50,000)									
Annual Payments on Tax-Exempt ESCO Loan	$ (196,896)	$ (196,896)	$ (196,896)	$ (196,896)	$ (196,896)	$ (196,896)	$ (196,896)	$ (196,896)	$ (196,896)	$ (196,896)
SAVINGS:										
Annual Savings	$ 350,000	$ 350,000	$ 350,000	$ 350,000	$ 350,000	$ 350,000	$ 350,000	$ 350,000	$ 350,000	$ 350,000
Net Annual Cash Flows:	$ 103,104	$ 153,104	$ 153,104	$ 153,104	$ 153,104	$ 153,104	$ 153,104	$ 153,104	$ 153,104	$ 153,104
PV's, Present Values:	$95,467	$131,262	$121,539	$112,536	$104,200	$96,481	$89,335	$82,717	$76,590	$70,917
NPV (1)	$981,044	Sum of ALL Present Values from the ten Year Life of the Equipment or Project								
NPV (2)	$981,044	Excel Based Calculation								
IRR	9%	Average of ALL Cash Flows, or Net Savings Generated By the Project, During its Ten Year Life Expectancy Divided by the Total Investment								
ROI:	9%	Net Inc./Avg.Total Assets:								
Payback Period, in YEARS:	10.97	Total Investment Divided By the Average of Cash Flows								

Energy Project Cost and Savings Breakdown			
No.	**Description**	**Measure Cost**	**Annual Savings**
1	Lighting System Upgrade	$900,000	$200,000
2	Installation of VFD's	$100,000	$25,000
3	Automated HVAC Control & EMS Systems	$400,000	$40,000
4	Replacement of Cooling Tower	$100,000	$10,000
5	Replacement of a chiller	$300,000	$40,000
6	Expansion of Existing Chilled Water System.	$300,000	$35,000
	Sub-Totals	**$2,100,000**	**$350,000**
	20% ESCO Discount.Cost Advantage	**$420,000.0**	
	Net Investment	**$1,680,000.0**	

Spread Sheet 11.4. Financial Analysis Of Investment By ESCO On Tax-Exempt Lease-Purchase Energy/Utility Project
- High Return Measures Only. Scenario II-B.

II - B. Project Financing Through Cash Borrowed from Commercial Financial Institution at a Lower Tax-Exempt Rate
Asssumed Annual Interest Rate for Tax-Exempt Lease-Purchase Financing of the Project: 3%
Assumed Annual Discount (Interest) Rate: 8%
Annual Payment Toward Amortization of Borrowed Capital = $800,000 (A/P, 3%,10) = ($800,000). (0.1172) = $ 93,760.00

Initial Cost of Project: $ 800,000	Project/Equipmnet Life: 10				Operating Years & Cash Flows					
Description:	1	2	3	4	5	6	7	8	9	10
COSTS:										
Write-off of Existing Investment, at Book Value	$ (50,000)									
Annual Payments on Tax-Exempt ESCO Loan	$ (93,760)	$ (93,760)	$ (93,760)	$ (93,760)	$ (93,760)	$ (93,760)	$ (93,760)	$ (93,760)	$ (93,760)	$ (93,760)
SAVINGS:										
Annual Savings	$ 225,000	$ 225,000	$ 225,000	$ 225,000	$ 225,000	$ 225,000	$ 225,000	$ 225,000	$ 225,000	$ 225,000
Net Annual Cash Flows:	$ 81,240	$ 131,240	$ 131,240	$ 131,240	$ 131,240	$ 131,240	$ 131,240	$ 131,240	$ 131,240	$ 131,240
PV's, Present Values:	$75,222	$112,517	$104,183	$96,465	$89,320	$82,703	$76,577	$70,905	$65,653	$60,790
NPV (1)	$834,335	Sum of ALL Present Values from the ten Year Life of the Equipment or Project								
NPV (2)	$834,335	Excel Based Calculation								
IRR	16%	Average of ALL Cash Flows, or Net Savings Generated By the Project, During its Ten Year Life Expectancy Divided by the Total Investment								
ROI:	16%	Net Inc./Avg.Total Assets:								
Payback Period, in YEARS:	6.10	Total Investment Divided By the Average of Cash Flows								

Energy Project Cost and Savings Breakdown			
No.	Description	Measure Cost	Annual Savings
1	Lighting System Upgrade	$900,000	$200,000
2	Installation of VFD's	$100,000	$25,000
	Sub-Totals	**$1,000,000**	**$225,000**
	20% ESCO Discount,Cost Advantage	**$200,000.0**	
	Net Investment	**$800,000.0**	

Spread Sheet 11.5. Financial Analysis Of Investment By ESCO On Tax-Exempt Lease-Purchase Energy/Utility Project

Low Return Measures Only. Scenario II-C.

II - C. Project Financing Through Cash Borrowed from Commercial Financial Institution at a Lower Tax-Exempt Rate
Assumed Annual Interest Rate for Tax-Exempt Lease-Purchase Financing of the Project: 3%
Assumed Annual Discount (Interest) Rate: 8%
Annual Payment Toward Amortization of Borrowed Capital = $880,000 (A/P, 3%,10) = ($880,000). (0.1172) = $ 103,136.00

Initial Cost of Project: $ 880,000	Project/Equipment Life: 10									
	Operating Years & Cash Flows									
Description:	1	2	3	4	5	6	7	8	9	10
COSTS: Write-off of Existing Investment, at Book Value	$ (50,000)									
Annual Payments on Tax-Exempt ESCO Loan	$ (103,136)	$ (103,136)	$ (103,136)	$ (103,136)	$ (103,136)	$ (103,136)	$ (103,136)	$ (103,136)	$ (103,136)	$ (103,136)
SAVINGS: Annual Savings	$ 125,000	$ 125,000	$ 125,000	$ 125,000	$ 125,000	$ 125,000	$ 125,000	$ 125,000	$ 125,000	$ 125,000
Net Annual Cash Flows:	$ (28,136)	$ 21,864	$ 21,864	$ 21,864	$ 21,864	$ 21,864	$ 21,864	$ 21,864	$ 21,864	$ 21,864
PV's, Present Values:	($26,052)	$18,745	$17,356	$16,071	$14,880	$13,778	$12,757	$11,812	$10,937	$10,127
NPV (1)	$100,413	Sum of ALL Present Values from the ten Year Life of the Equipment or Project								
NPV (2)	$100,413	Excel Based Calculation								
IRR	2%	Average of ALL Cash Flows, or Net Savings Generated By the Project, During its Ten Year Life Expectancy Divided by the Total Investment								
ROI:	2%	Net Inc./Avg.Total Assets:								
Payback Period, in YEARS:	40.25	Total Investment Divided By the Average of Cash Flows								

Energy Project Cost and Savings Breakdown

No.	Description	Measure Cost	Annual Savings
1	Automated HVAC Control & EMS Systems	$400,000	$40,000
2	Replacement of Cooling Tower	$100,000	$10,000
3	Replacement of a chiller	$300,000	$40,000
4	Expansion of Existing Chilled Water System.	$300,000	$35,000
	Sub-Totals	**$1,100,000**	**$125,000**
	20% ESCO Discount Cost Advantage	**$220,000.0**	
	Net Investment	**$880,000.0**	

Appendix C

Financial Factors for Time Value of Money Calculations

Financial Factors for Interest Rate, i = 0.5%

N	(P/F)	(P/A)	(F/P)	(F/A)	(A/P)	(A/F)	(P/G)	(A/G)	(F/G)
1	0.9950	0.9950	1.0050	1.0000	1.0050	1.0000	0.0000	0.0000	0.0000
2	0.9901	1.9851	1.0100	2.0050	0.5038	0.4988	0.9901	0.4988	1.0000
3	0.9851	2.9702	1.0151	3.0150	0.3367	0.3317	2.9604	0.9967	3.0050
4	0.9802	3.9505	1.0202	4.0301	0.2531	0.2481	5.9011	1.4938	6.0200
5	0.9754	4.9259	1.0253	5.0503	0.2030	0.1980	9.8026	1.9900	10.0501
6	0.9705	5.8964	1.0304	6.0755	0.1696	0.1646	14.6552	2.4855	15.1004
7	0.9657	6.8621	1.0355	7.1059	0.1457	0.1407	20.4493	2.9801	21.1759
8	0.9609	7.8230	1.0407	8.1414	0.1278	0.1228	27.1755	3.4738	28.2818
9	0.9561	8.7791	1.0459	9.1821	0.1139	0.1089	34.8244	3.9668	36.4232
10	0.9513	9.7304	1.0511	10.2280	0.1028	0.0978	43.3865	4.4589	45.6053
11	0.9466	10.6770	1.0564	11.2792	0.0937	0.0887	52.8526	4.9501	55.8333
12	0.9419	11.6189	1.0617	12.3356	0.0861	0.0811	63.2136	5.4406	67.1125
13	0.9372	12.5562	1.0670	13.3972	0.0796	0.0746	74.4602	5.9302	79.4480
14	0.9326	13.4887	1.0723	14.4642	0.0741	0.0691	86.5835	6.4190	92.8453
15	0.9279	14.4166	1.0777	15.5365	0.0694	0.0644	99.5743	6.9069	107.3095
16	0.9233	15.3399	1.0831	16.6142	0.0652	0.0602	113.4238	7.3940	122.8461
17	0.9187	16.2586	1.0885	17.6973	0.0615	0.0565	128.1231	7.8803	139.4603
18	0.9141	17.1728	1.0939	18.7858	0.0582	0.0532	143.6634	8.3658	157.1576
19	0.9096	18.0824	1.0994	19.8797	0.0553	0.0503	160.0360	8.8504	175.9434
20	0.9051	18.9874	1.1049	20.9791	0.0527	0.0477	177.2322	9.3342	195.8231
21	0.9006	19.8880	1.1104	22.0840	0.0503	0.0453	195.2434	9.8172	216.8022
22	0.8961	20.7841	1.1160	23.1944	0.0481	0.0431	214.0611	10.2993	238.8862
23	0.8916	21.6757	1.1216	24.3104	0.0461	0.0411	233.6768	10.7806	262.0806
24	0.8872	22.5629	1.1272	25.4320	0.0443	0.0393	254.0820	11.2611	286.3910
25	0.8828	23.4456	1.1328	26.5591	0.0427	0.0377	275.2686	11.7407	311.8230
26	0.8784	24.3240	1.1385	27.6919	0.0411	0.0361	297.2281	12.2195	338.3821
27	0.8740	25.1980	1.1442	28.8304	0.0397	0.0347	319.9523	12.6975	366.0740
28	0.8697	26.0677	1.1499	29.9745	0.0384	0.0334	343.4332	13.1747	394.9044
29	0.8653	26.9330	1.1556	31.1244	0.0371	0.0321	367.6625	13.6510	424.8789
30	0.8610	27.7941	1.1614	32.2800	0.0360	0.0310	392.6324	14.1265	456.0033
31	0.8567	28.6508	1.1672	33.4414	0.0349	0.0299	418.3348	14.6012	488.2833
32	0.8525	29.5033	1.1730	34.6086	0.0339	0.0289	444.7618	15.0750	521.7247
33	0.8482	30.3515	1.1789	35.7817	0.0329	0.0279	471.9055	15.5480	556.3334
34	0.8440	31.1955	1.1848	36.9606	0.0321	0.0271	499.7583	16.0202	592.1150
35	0.8398	32.0354	1.1907	38.1454	0.0312	0.0262	528.3123	16.4915	629.0756
36	0.8356	32.8710	1.1967	39.3361	0.0304	0.0254	557.5598	16.9621	667.2210
37	0.8315	33.7025	1.2027	40.5328	0.0297	0.0247	587.4934	17.4317	706.5571

Financial Factors for Interest Rate, i = 0.5%

N	(P/F)	(P/A)	(F/P)	(F/A)	(A/P)	(A/F)	(P/G)	(A/G)	(F/G)
38	0.8274	34.5299	1.2087	41.7354	0.0290	0.0240	618.1054	17.9006	747.0899
39	0.8232	35.3531	1.2147	42.9441	0.0283	0.0233	649.3883	18.3686	788.8253
40	0.8191	36.1722	1.2208	44.1588	0.0276	0.0226	681.3347	18.8359	831.7695
41	0.8151	36.9873	1.2269	45.3796	0.0270	0.0220	713.9372	19.3022	875.9283
42	0.8110	37.7983	1.2330	46.6065	0.0265	0.0215	747.1886	19.7678	921.3079
43	0.8070	38.6053	1.2392	47.8396	0.0259	0.0209	781.0815	20.2325	967.9145
44	0.8030	39.4082	1.2454	49.0788	0.0254	0.0204	815.6087	20.6964	1015.7541
45	0.7990	40.2072	1.2516	50.3242	0.0249	0.0199	850.7631	21.1595	1064.8328
46	0.7950	41.0022	1.2579	51.5758	0.0244	0.0194	886.5376	21.6217	1115.1570
47	0.7910	41.7932	1.2642	52.8337	0.0239	0.0189	922.9252	22.0831	1166.7328
48	0.7871	42.5803	1.2705	54.0978	0.0235	0.0185	959.9188	22.5437	1219.5664
49	0.7832	43.3635	1.2768	55.3683	0.0231	0.0181	997.5116	23.0035	1273.6643
50	0.7793	44.1428	1.2832	56.6452	0.0227	0.0177	1035.6966	23.4624	1329.0326
51	0.7754	44.9182	1.2896	57.9284	0.0223	0.0173	1074.4670	23.9205	1385.6778
52	0.7716	45.6897	1.2961	59.2180	0.0219	0.0169	1113.8162	24.3778	1443.6061
53	0.7677	46.4575	1.3026	60.5141	0.0215	0.0165	1153.7372	24.8343	1502.8242
54	0.7639	47.2214	1.3091	61.8167	0.0212	0.0162	1194.2236	25.2899	1563.3383
55	0.7601	47.9814	1.3156	63.1258	0.0208	0.0158	1235.2686	25.7447	1625.1550
56	0.7563	48.7378	1.3222	64.4414	0.0205	0.0155	1276.8657	26.1987	1688.2808
57	0.7525	49.4903	1.3288	65.7636	0.0202	0.0152	1319.0084	26.6519	1752.7222
58	0.7488	50.2391	1.3355	67.0924	0.0199	0.0149	1361.6903	27.1042	1818.4858
59	0.7451	50.9842	1.3421	68.4279	0.0196	0.0146	1404.9048	27.5557	1885.5782
60	0.7414	51.7256	1.3489	69.7700	0.0193	0.0143	1448.6458	28.0064	1954.0061
65	0.7231	55.3775	1.3829	76.5821	0.0181	0.0131	1675.0272	30.2475	2316.4124
70	0.7053	58.9394	1.4178	83.5661	0.0170	0.0120	1913.6427	32.4680	2713.2211
75	0.6879	62.4136	1.4536	90.7265	0.0160	0.0110	2163.7525	34.6679	3145.3010
80	0.6710	65.8023	1.4903	98.0677	0.0152	0.0102	2424.6455	36.8474	3613.5427
85	0.6545	69.1075	1.5280	105.5943	0.0145	0.0095	2695.6389	39.0065	4118.8594
90	0.6383	72.3313	1.5666	113.3109	0.0138	0.0088	2976.0769	41.1451	4662.1872
95	0.6226	75.4757	1.6061	121.2224	0.0132	0.0082	3265.3298	43.2633	5244.4859
100	0.6073	78.5426	1.6467	129.3337	0.0127	0.0077	3562.7934	45.3613	5866.7397

Financial Factors, Interest Rate, i = 1%

N	(P/F)	(P/A)	(F/P)	(F/A)	(A/P)	(A/F)	(P/G)	(A/G)	(F/G)
1	0.9901	0.9901	1.0100	1.0000	1.0100	1.0000	0.0000	0.0000	0.0000
2	0.9803	1.9704	1.0201	2.0100	0.5075	0.4975	0.9803	0.4975	1.0000
3	0.9706	2.9410	1.0303	3.0301	0.3400	0.3300	2.9215	0.9934	3.0100
4	0.9610	3.9020	1.0406	4.0604	0.2563	0.2463	5.8044	1.4876	6.0401
5	0.9515	4.8534	1.0510	5.1010	0.2060	0.1960	9.6103	1.9801	10.1005
6	0.9420	5.7955	1.0615	6.1520	0.1725	0.1625	14.3205	2.4710	15.2015
7	0.9327	6.7282	1.0721	7.2135	0.1486	0.1386	19.9168	2.9602	21.3535
8	0.9235	7.6517	1.0829	8.2857	0.1307	0.1207	26.3812	3.4478	28.5671
9	0.9143	8.5660	1.0937	9.3685	0.1167	0.1067	33.6959	3.9337	36.8527
10	0.9053	9.4713	1.1046	10.4622	0.1056	0.0956	41.8435	4.4179	46.2213
11	0.8963	10.3676	1.1157	11.5668	0.0965	0.0865	50.8067	4.9005	56.6835
12	0.8874	11.2551	1.1268	12.6825	0.0888	0.0788	60.5687	5.3815	68.2503
13	0.8787	12.1337	1.1381	13.8093	0.0824	0.0724	71.1126	5.8607	80.9328
14	0.8700	13.0037	1.1495	14.9474	0.0769	0.0669	82.4221	6.3384	94.7421
15	0.8613	13.8651	1.1610	16.0969	0.0721	0.0621	94.4810	6.8143	109.6896
16	0.8528	14.7179	1.1726	17.2579	0.0679	0.0579	107.2734	7.2886	125.7864

Financial Factors for Interest Rate, i = 1%

N	(P/F)	(P/A)	(F/P)	(F/A)	(A/P)	(A/F)	(P/G)	(A/G)	(F/G)
17	0.8444	15.5623	1.1843	18.4304	0.0643	0.0543	120.7834	7.7613	143.0443
18	0.8360	16.3983	1.1961	19.6147	0.0610	0.0510	134.9957	8.2323	161.4748
19	0.8277	17.2260	1.2081	20.8109	0.0581	0.0481	149.8950	8.7017	181.0895
20	0.8195	18.0456	1.2202	22.0190	0.0554	0.0454	165.4664	9.1694	201.9004
21	0.8114	18.8570	1.2324	23.2392	0.0530	0.0430	181.6950	9.6354	223.9194
22	0.8034	19.6604	1.2447	24.4716	0.0509	0.0409	198.5663	10.0998	247.1586
23	0.7954	20.4558	1.2572	25.7163	0.0489	0.0389	216.0660	10.5626	271.6302
24	0.7876	21.2434	1.2697	26.9735	0.0471	0.0371	234.1800	11.0237	297.3465
25	0.7798	22.0232	1.2824	28.2432	0.0454	0.0354	252.8945	11.4831	324.3200
26	0.7720	22.7952	1.2953	29.5256	0.0439	0.0339	272.1957	11.9409	352.5631
27	0.7644	23.5596	1.3082	30.8209	0.0424	0.0324	292.0702	12.3971	382.0888
28	0.7568	24.3164	1.3213	32.1291	0.0411	0.0311	312.5047	12.8516	412.9097
29	0.7493	25.0658	1.3345	33.4504	0.0399	0.0299	333.4863	13.3044	445.0388
30	0.7419	25.8077	1.3478	34.7849	0.0387	0.0287	355.0021	13.7557	478.4892
31	0.7346	26.5423	1.3613	36.1327	0.0377	0.0277	377.0394	14.2052	513.2740
32	0.7273	27.2696	1.3749	37.4941	0.0367	0.0267	399.5858	14.6532	549.4068
33	0.7201	27.9897	1.3887	38.8690	0.0357	0.0257	422.6291	15.0995	586.9009
34	0.7130	28.7027	1.4026	40.2577	0.0348	0.0248	446.1572	15.5441	625.7699
35	0.7059	29.4086	1.4166	41.6603	0.0340	0.0240	470.1583	15.9871	666.0276
36	0.6989	30.1075	1.4308	43.0769	0.0332	0.0232	494.6207	16.4285	707.6878
37	0.6920	30.7995	1.4451	44.5076	0.0325	0.0225	519.5329	16.8682	750.7647
38	0.6852	31.4847	1.4595	45.9527	0.0318	0.0218	544.8835	17.3063	795.2724
39	0.6784	32.1630	1.4741	47.4123	0.0311	0.0211	570.6616	17.7428	841.2251
40	0.6717	32.8347	1.4889	48.8864	0.0305	0.0205	596.8561	18.1776	888.6373
41	0.6650	33.4997	1.5038	50.3752	0.0299	0.0199	623.4562	18.6108	937.5237
42	0.6584	34.1581	1.5188	51.8790	0.0293	0.0193	650.4514	19.0424	987.8989
43	0.6519	34.8100	1.5340	53.3978	0.0287	0.0187	677.8312	19.4723	1039.7779
44	0.6454	35.4555	1.5493	54.9318	0.0282	0.0182	705.5853	19.9006	1093.1757
45	0.6391	36.0945	1.5648	56.4811	0.0277	0.0177	733.7037	20.3273	1148.1075
46	0.6327	36.7272	1.5805	58.0459	0.0272	0.0172	762.1765	20.7524	1204.5885
47	0.6265	37.3537	1.5963	59.6263	0.0268	0.0168	790.9938	21.1758	1262.6344
48	0.6203	37.9740	1.6122	61.2226	0.0263	0.0163	820.1460	21.5976	1322.2608
49	0.6141	38.5881	1.6283	62.8348	0.0259	0.0159	849.6237	22.0178	1383.4834
50	0.6080	39.1961	1.6446	64.4632	0.0255	0.0155	879.4176	22.4363	1446.3182
51	0.6020	39.7981	1.6611	66.1078	0.0251	0.0151	909.5186	22.8533	1510.7814
52	0.5961	40.3942	1.6777	67.7689	0.0248	0.0148	939.9175	23.2686	1576.8892
53	0.5902	40.9844	1.6945	69.4466	0.0244	0.0144	970.6057	23.6823	1644.6581
54	0.5843	41.5687	1.7114	71.1410	0.0241	0.0141	1001.5743	24.0945	1714.1047
55	0.5785	42.1472	1.7285	72.8525	0.0237	0.0137	1032.8148	24.5049	1785.2457
56	0.5728	42.7200	1.7458	74.5810	0.0234	0.0134	1064.3188	24.9138	1858.0982
57	0.5671	43.2871	1.7633	76.3268	0.0231	0.0131	1096.0780	25.3211	1932.6792
58	0.5615	43.8486	1.7809	78.0901	0.0228	0.0128	1128.0843	25.7268	2009.0060
59	0.5560	44.4046	1.7987	79.8710	0.0225	0.0125	1160.3296	26.1308	2087.0960
60	0.5504	44.9550	1.8167	81.6697	0.0222	0.0122	1192.8061	26.5333	2166.9670
65	0.5237	47.6266	1.9094	90.9366	0.0210	0.0110	1358.3903	28.5217	2593.6649
70	0.4983	50.1685	2.0068	100.6763	0.0199	0.0099	1528.6474	30.4703	3067.6337
75	0.4741	52.5871	2.1091	110.9128	0.0190	0.0090	1702.7340	32.3793	3591.2847
80	0.4511	54.8882	2.2167	121.6715	0.0182	0.0082	1879.8771	34.2492	4167.1522
85	0.4292	57.0777	2.3298	132.9790	0.0175	0.0075	2059.3701	36.0801	4797.8997
90	0.4084	59.1609	2.4486	144.8633	0.0169	0.0069	2240.5675	37.8724	5486.3267
95	0.3886	61.1430	2.5735	157.3538	0.0164	0.0064	2422.8811	39.6265	6235.3755
100	0.3697	63.0289	2.7048	170.4814	0.0159	0.0059	2605.7758	41.3426	7048.1383

Financial Factors, Interest Rate, i = 2%

N	(P/F)	(P/A)	(F/P)	(F/A)	(A/P)	(A/F)	(P/G)	(A/G)	(F/G)
1	0.9804	0.9804	1.0200	1.0000	1.0200	1.0000	0.0000	0.0000	0.0000
2	0.9612	1.9416	1.0404	2.0200	0.5150	0.4950	0.9612	0.4950	1.0000
3	0.9423	2.8839	1.0612	3.0604	0.3468	0.3268	2.8458	0.9868	3.0200
4	0.9238	3.8077	1.0824	4.1216	0.2626	0.2426	5.6173	1.4752	6.0804
5	0.9057	4.7135	1.1041	5.2040	0.2122	0.1922	9.2403	1.9604	10.2020
6	0.8880	5.6014	1.1262	6.3081	0.1785	0.1585	13.6801	2.4423	15.4060
7	0.8706	6.4720	1.1487	7.4343	0.1545	0.1345	18.9035	2.9208	21.7142
8	0.8535	7.3255	1.1717	8.5830	0.1365	0.1165	24.8779	3.3961	29.1485
9	0.8368	8.1622	1.1951	9.7546	0.1225	0.1025	31.5720	3.8681	37.7314
10	0.8203	8.9826	1.2190	10.9497	0.1113	0.0913	38.9551	4.3367	47.4860
11	0.8043	9.7868	1.2434	12.1687	0.1022	0.0822	46.9977	4.8021	58.4358
12	0.7885	10.5753	1.2682	13.4121	0.0946	0.0746	55.6712	5.2642	70.6045
13	0.7730	11.3484	1.2936	14.6803	0.0881	0.0681	64.9475	5.7231	84.0166
14	0.7579	12.1062	1.3195	15.9739	0.0826	0.0626	74.7999	6.1786	98.6969
15	0.7430	12.8493	1.3459	17.2934	0.0778	0.0578	85.2021	6.6309	114.6708
16	0.7284	13.5777	1.3728	18.6393	0.0737	0.0537	96.1288	7.0799	131.9643
17	0.7142	14.2919	1.4002	20.0121	0.0700	0.0500	107.5554	7.5256	150.6035
18	0.7002	14.9920	1.4282	21.4123	0.0667	0.0467	119.4581	7.9681	170.6156
19	0.6864	15.6785	1.4568	22.8406	0.0638	0.0438	131.8139	8.4073	192.0279
20	0.6730	16.3514	1.4859	24.2974	0.0612	0.0412	144.6003	8.8433	214.8685
21	0.6598	17.0112	1.5157	25.7833	0.0588	0.0388	157.7959	9.2760	239.1659
22	0.6468	17.6580	1.5460	27.2990	0.0566	0.0366	171.3795	9.7055	264.9492
23	0.6342	18.2922	1.5769	28.8450	0.0547	0.0347	185.3309	10.1317	292.2482
24	0.6217	18.9139	1.6084	30.4219	0.0529	0.0329	199.6305	10.5547	321.0931
25	0.6095	19.5235	1.6406	32.0303	0.0512	0.0312	214.2592	10.9745	351.5150
26	0.5976	20.1210	1.6734	33.6709	0.0497	0.0297	229.1987	11.3910	383.5453
27	0.5859	20.7069	1.7069	35.3443	0.0483	0.0283	244.4311	11.8043	417.2162
28	0.5744	21.2813	1.7410	37.0512	0.0470	0.0270	259.9392	12.2145	452.5605
29	0.5631	21.8444	1.7758	38.7922	0.0458	0.0258	275.7064	12.6214	489.6117
30	0.5521	22.3965	1.8114	40.5681	0.0446	0.0246	291.7164	13.0251	528.4040
31	0.5412	22.9377	1.8476	42.3794	0.0436	0.0236	307.9538	13.4257	568.9720
32	0.5306	23.4683	1.8845	44.2270	0.0426	0.0226	324.4035	13.8230	611.3515
33	0.5202	23.9886	1.9222	46.1116	0.0417	0.0217	341.0508	14.2172	655.5785
34	0.5100	24.4986	1.9607	48.0338	0.0408	0.0208	357.8817	14.6083	701.6901
35	0.5000	24.9986	1.9999	49.9945	0.0400	0.0200	374.8826	14.9961	749.7239
36	0.4902	25.4888	2.0399	51.9944	0.0392	0.0192	392.0405	15.3809	799.7184
37	0.4806	25.9695	2.0807	54.0343	0.0385	0.0185	409.3424	15.7625	851.713
38	0.4712	26.4406	2.1223	56.1149	0.0378	0.0178	426.7764	16.1409	905.747
39	0.4619	26.9026	2.1647	58.2372	0.0372	0.0172	444.3304	16.5163	961.862
40	0.4529	27.3555	2.2080	60.4020	0.0366	0.0166	461.9931	16.8885	1020.099
41	0.4440	27.7995	2.2522	62.6100	0.0360	0.0160	479.7535	17.2576	1080.501
42	0.4353	28.2348	2.2972	64.8622	0.0354	0.0154	497.6010	17.6237	1143.111
43	0.4268	28.6616	2.3432	67.1595	0.0349	0.0149	515.5253	17.9866	1207.973
44	0.4184	29.0800	2.3901	69.5027	0.0344	0.0144	533.5165	18.3465	1275.133
45	0.4102	29.4902	2.4379	71.8927	0.0339	0.0139	551.5652	18.7034	1344.636
46	0.4022	29.8923	2.4866	74.3306	0.0335	0.0135	569.6621	19.0571	1416.528
47	0.3943	30.2866	2.5363	76.8172	0.0330	0.0130	587.7985	19.4079	1490.859
48	0.3865	30.6731	2.5871	79.3535	0.0326	0.0126	605.9657	19.7556	1567.676
49	0.3790	31.0521	2.6388	81.9406	0.0322	0.0122	624.1557	20.1003	1647.029
50	0.3715	31.4236	2.6916	84.5794	0.0318	0.0118	642.3606	20.4420	1728.970
51	0.3642	31.7878	2.7454	87.2710	0.0315	0.0115	660.5727	20.7807	1813.549
52	0.3571	32.1449	2.8003	90.0164	0.0311	0.0111	678.7849	21.1164	1900.820
53	0.3501	32.4950	2.8563	92.8167	0.0308	0.0108	696.9900	21.4491	1990.837
54	0.3432	32.8383	2.9135	95.6731	0.0305	0.0105	715.1815	21.7789	2083.654

Financial Factors for Interest Rate, i = 2%

N	(P/F)	(P/A)	(F/P)	(F/A)	(A/P)	(A/F)	(P/G)	(A/G)	(F/G)
55	0.3365	33.1748	2.9717	98.5865	0.0301	0.0101	733.3527	22.1057	2179.327
56	0.3299	33.5047	3.0312	101.5583	0.0298	0.0098	751.4975	22.4296	2277.913
57	0.3234	33.8281	3.0918	104.5894	0.0296	0.0096	769.6100	22.7506	2379.471
58	0.3171	34.1452	3.1536	107.6812	0.0293	0.0093	787.6845	23.0687	2484.061
59	0.3109	34.4561	3.2167	110.8348	0.0290	0.0090	805.7154	23.3838	2591.742
60	0.3048	34.7609	3.2810	114.0515	0.0288	0.0088	823.6975	23.6961	2702.577
65	0.2761	36.1975	3.6225	131.1262	0.0276	0.0076	912.7085	25.2147	3306.308
70	0.2500	37.4986	3.9996	149.9779	0.0267	0.0067	999.8343	26.6632	3998.896
75	0.2265	38.6771	4.4158	170.7918	0.0259	0.0059	1084.6393	28.0434	4789.589
80	0.2051	39.7445	4.8754	193.7720	0.0252	0.0052	1166.7868	29.3572	5688.598
85	0.1858	40.7113	5.3829	219.1439	0.0246	0.0046	1246.0241	30.6064	6707.197
90	0.1683	41.5869	5.9431	247.1567	0.0240	0.0040	1322.1701	31.7929	7857.833
95	0.1524	42.3800	6.5617	278.0850	0.0236	0.0036	1395.1033	32.9189	9154.248
100	0.1380	43.0984	7.2446	312.2323	0.0232	0.0032	1464.7527	33.9863	10611.615

Financial Factors, Interest Rate, i = 3%

N	(P/F)	(P/A)	(F/P)	(F/A)	(A/P)	(A/F)	(P/G)	(A/G)	(F/G)
1	0.9709	0.9709	1.0300	1.0000	1.0300	1.0000	0.0000	0.0000	0.0000
2	0.9426	1.9135	1.0609	2.0300	0.5226	0.4926	0.9426	0.4926	1.0000
3	0.9151	2.8286	1.0927	3.0909	0.3535	0.3235	2.7729	0.9803	3.0300
4	0.8885	3.7171	1.1255	4.1836	0.2690	0.2390	5.4383	1.4631	6.1209
5	0.8626	4.5797	1.1593	5.3091	0.2184	0.1884	8.8888	1.9409	10.3045
6	0.8375	5.4172	1.1941	6.4684	0.1846	0.1546	13.0762	2.4138	15.6137
7	0.8131	6.2303	1.2299	7.6625	0.1605	0.1305	17.9547	2.8819	22.0821
8	0.7894	7.0197	1.2668	8.8923	0.1425	0.1125	23.4806	3.3450	29.7445
9	0.7664	7.7861	1.3048	10.1591	0.1284	0.0984	29.6119	3.8032	38.6369
10	0.7441	8.5302	1.3439	11.4639	0.1172	0.0872	36.3088	4.2565	48.7960
11	0.7224	9.2526	1.3842	12.8078	0.1081	0.0781	43.5330	4.7049	60.2599
12	0.7014	9.9540	1.4258	14.1920	0.1005	0.0705	51.2482	5.1485	73.0677
13	0.6810	10.6350	1.4685	15.6178	0.0940	0.0640	59.4196	5.5872	87.2597
14	0.6611	11.2961	1.5126	17.0863	0.0885	0.0585	68.0141	6.0210	102.8775
15	0.6419	11.9379	1.5580	18.5989	0.0838	0.0538	77.0002	6.4500	119.9638
16	0.6232	12.5611	1.6047	20.1569	0.0796	0.0496	86.3477	6.8742	138.5627
17	0.6050	13.1661	1.6528	21.7616	0.0760	0.0460	96.0280	7.2936	158.7196
18	0.5874	13.7535	1.7024	23.4144	0.0727	0.0427	106.0137	7.7081	180.4812
19	0.5703	14.3238	1.7535	25.1169	0.0698	0.0398	116.2788	8.1179	203.8956
20	0.5537	14.8775	1.8061	26.8704	0.0672	0.0372	126.7987	8.5229	229.0125
21	0.5375	15.4150	1.8603	28.6765	0.0649	0.0349	137.5496	8.9231	255.8829
22	0.5219	15.9369	1.9161	30.5368	0.0627	0.0327	148.5094	9.3186	284.5593
23	0.5067	16.4436	1.9736	32.4529	0.0608	0.0308	159.6566	9.7093	315.0961
24	0.4919	16.9355	2.0328	34.4265	0.0590	0.0290	170.9711	10.0954	347.5490
25	0.4776	17.4131	2.0938	36.4593	0.0574	0.0274	182.4336	10.4768	381.9755
26	0.4637	17.8768	2.1566	38.5530	0.0559	0.0259	194.0260	10.8535	418.4347
27	0.4502	18.3270	2.2213	40.7096	0.0546	0.0246	205.7309	11.2255	456.9878
28	0.4371	18.7641	2.2879	42.9309	0.0533	0.0233	217.5320	11.5930	497.6974
29	0.4243	19.1885	2.3566	45.2189	0.0521	0.0221	229.4137	11.9558	540.6283
30	0.4120	19.6004	2.4273	47.5754	0.0510	0.0210	241.3613	12.3141	585.8472
31	0.4000	20.0004	2.5001	50.0027	0.0500	0.0200	253.3609	12.6678	633.4226
32	0.3883	20.3888	2.5751	52.5028	0.0490	0.0190	265.3993	13.0169	683.4253

Financial Factors, Interest Rate, i = 3%

N	(P/F)	(P/A)	(F/P)	(F/A)	(A/P)	(A/F)	(P/G)	(A/G)	(F/G)
33	0.3770	20.7658	2.6523	55.0778	0.0482	0.0182	277.4642	13.3616	735.9280
34	0.3660	21.1318	2.7319	57.7302	0.0473	0.0173	289.5437	13.7018	791.0059
35	0.3554	21.4872	2.8139	60.4621	0.0465	0.0165	301.6267	14.0375	848.7361
36	0.3450	21.8323	2.8983	63.2759	0.0458	0.0158	313.7028	14.3688	909.1981
37	0.3350	22.1672	2.9852	66.1742	0.0451	0.0151	325.7622	14.6957	972.4741
38	0.3252	22.4925	3.0748	69.1594	0.0445	0.0145	337.7956	15.0182	1038.6483
39	0.3158	22.8082	3.1670	72.2342	0.0438	0.0138	349.7942	15.3363	1107.8078
40	0.3066	23.1148	3.2620	75.4013	0.0433	0.0133	361.7499	15.6502	1180.0420
41	0.2976	23.4124	3.3599	78.6633	0.0427	0.0127	373.6551	15.9597	1255.4433
42	0.2890	23.7014	3.4607	82.0232	0.0422	0.0122	385.5024	16.2650	1334.1065
43	0.2805	23.9819	3.5645	85.4839	0.0417	0.0117	397.2852	16.5660	1416.1297
44	0.2724	24.2543	3.6715	89.0484	0.0412	0.0112	408.9972	16.8629	1501.6136
45	0.2644	24.5187	3.7816	92.7199	0.0408	0.0108	420.6325	17.1556	1590.6620
46	0.2567	24.7754	3.8950	96.5015	0.0404	0.0104	432.1856	17.4441	1683.3819
47	0.2493	25.0247	4.0119	100.3965	0.0400	0.0100	443.6515	17.7285	1779.8834
48	0.2420	25.2667	4.1323	104.4084	0.0396	0.0096	455.0255	18.0089	1880.2799
49	0.2350	25.5017	4.2562	108.5406	0.0392	0.0092	466.3031	18.2852	1984.6883
50	0.2281	25.7298	4.3839	112.7969	0.0389	0.0089	477.4803	18.5575	2093.2289
51	0.2215	25.9512	4.5154	117.1808	0.0385	0.0085	488.5535	18.8258	2206.0258
52	0.2150	26.1662	4.6509	121.6962	0.0382	0.0082	499.5191	19.0902	2323.2066
53	0.2088	26.3750	4.7904	126.3471	0.0379	0.0079	510.3742	19.3507	2444.9027
54	0.2027	26.5777	4.9341	131.1375	0.0376	0.0076	521.1157	19.6073	2571.2498
55	0.1968	26.7744	5.0821	136.0716	0.0373	0.0073	531.7411	19.8600	2702.3873
56	0.1910	26.9655	5.2346	141.1538	0.0371	0.0071	542.2481	20.1090	2838.4589
57	0.1855	27.1509	5.3917	146.3884	0.0368	0.0068	552.6345	20.3542	2979.6127
58	0.1801	27.3310	5.5534	151.7800	0.0366	0.0066	562.8985	20.5956	3126.0011
59	0.1748	27.5058	5.7200	157.3334	0.0364	0.0064	573.0384	20.8333	3277.7811
60	0.1697	27.6756	5.8916	163.0534	0.0361	0.0061	583.0526	21.0674	3435.1146
65	0.1464	28.4529	6.8300	194.3328	0.0351	0.0051	631.2010	22.1841	4311.0919
70	0.1263	29.1234	7.9178	230.5941	0.0343	0.0043	676.0869	23.2145	5353.1355
75	0.1089	29.7018	9.1789	272.6309	0.0337	0.0037	717.6978	24.1634	6587.6952
80	0.0940	30.2008	10.6409	321.3630	0.0331	0.0031	756.0865	25.0353	8045.4340
85	0.0811	30.6312	12.3357	377.8570	0.0326	0.0026	791.3529	25.8349	9761.8984
90	0.0699	31.0024	14.3005	443.3489	0.0323	0.0023	823.6302	26.5667	11778.2968
95	0.0603	31.3227	16.5782	519.2720	0.0319	0.0019	853.0742	27.2351	14142.4009
100	0.0520	31.5989	19.2186	607.2877	0.0316	0.0016	879.8540	27.8444	16909.5911

Financial Factors, Interest Rate, i = 4%

N	(P/F)	(P/A)	(F/P)	(F/A)	(A/P)	(A/F)	(P/G)	(A/G)	(F/G)
1	0.9615	0.9615	1.0400	1.0000	1.0400	1.0000	0.0000	0.0000	0.0000
2	0.9246	1.8861	1.0816	2.0400	0.5302	0.4902	0.9246	0.4902	1.0000
3	0.8890	2.7751	1.1249	3.1216	0.3603	0.3203	2.7025	0.9739	3.0400
4	0.8548	3.6299	1.1699	4.2465	0.2755	0.2355	5.2670	1.4510	6.1616
5	0.8219	4.4518	1.2167	5.4163	0.2246	0.1846	8.5547	1.9216	10.4081
6	0.7903	5.2421	1.2653	6.6330	0.1908	0.1508	12.5062	2.3857	15.8244
7	0.7599	6.0021	1.3159	7.8983	0.1666	0.1266	17.0657	2.8433	22.4574
8	0.7307	6.7327	1.3686	9.2142	0.1485	0.1085	22.1806	3.2944	30.3557
9	0.7026	7.4353	1.4233	10.5828	0.1345	0.0945	27.8013	3.7391	39.5699
10	0.6756	8.1109	1.4802	12.0061	0.1233	0.0833	33.8814	4.1773	50.1527

Financial Factors, Interest Rate, i =				4%					
N	(P/F)	(P/A)	(F/P)	(F/A)	(A/P)	(A/F)	(P/G)	(A/G)	(F/G)
11	0.6496	8.7605	1.5395	13.4864	0.1141	0.0741	40.3772	4.6090	62.1588
12	0.6246	9.3851	1.6010	15.0258	0.1066	0.0666	47.2477	5.0343	75.6451
13	0.6006	9.9856	1.6651	16.6268	0.1001	0.0601	54.4546	5.4533	90.6709
14	0.5775	10.5631	1.7317	18.2919	0.0947	0.0547	61.9618	5.8659	107.2978
15	0.5553	11.1184	1.8009	20.0236	0.0899	0.0499	69.7355	6.2721	125.5897
16	0.5339	11.6523	1.8730	21.8245	0.0858	0.0458	77.7441	6.6720	145.6133
17	0.5134	12.1657	1.9479	23.6975	0.0822	0.0422	85.9581	7.0656	167.4378
18	0.4936	12.6593	2.0258	25.6454	0.0790	0.0390	94.3498	7.4530	191.1353
19	0.4746	13.1339	2.1068	27.6712	0.0761	0.0361	102.8933	7.8342	216.7807
20	0.4564	13.5903	2.1911	29.7781	0.0736	0.0336	111.5647	8.2091	244.4520
21	0.4388	14.0292	2.2788	31.9692	0.0713	0.0313	120.3414	8.5779	274.2300
22	0.4220	14.4511	2.3699	34.2480	0.0692	0.0292	129.2024	8.9407	306.1992
23	0.4057	14.8568	2.4647	36.6179	0.0673	0.0273	138.1284	9.2973	340.4472
24	0.3901	15.2470	2.5633	39.0826	0.0656	0.0256	147.1012	9.6479	377.0651
25	0.3751	15.6221	2.6658	41.6459	0.0640	0.0240	156.1040	9.9925	416.1477
26	0.3607	15.9828	2.7725	44.3117	0.0626	0.0226	165.1212	10.3312	457.7936
27	0.3468	16.3296	2.8834	47.0842	0.0612	0.0212	174.1385	10.6640	502.1054
28	0.3335	16.6631	2.9987	49.9676	0.0600	0.0200	183.1424	10.9909	549.1896
29	0.3207	16.9837	3.1187	52.9663	0.0589	0.0189	192.1206	11.3120	599.1572
30	0.3083	17.2920	3.2434	56.0849	0.0578	0.0178	201.0618	11.6274	652.1234
31	0.2965	17.5885	3.3731	59.3283	0.0569	0.0169	209.9556	11.9371	708.2084
32	0.2851	17.8736	3.5081	62.7015	0.0559	0.0159	218.7924	12.2411	767.5367
33	0.2741	18.1476	3.6484	66.2095	0.0551	0.0151	227.5634	12.5396	830.2382
34	0.2636	18.4112	3.7943	69.8579	0.0543	0.0143	236.2607	12.8324	896.4477
35	0.2534	18.6646	3.9461	73.6522	0.0536	0.0136	244.8768	13.1198	966.3056
36	0.2437	18.9083	4.1039	77.5983	0.0529	0.0129	253.4052	13.4018	1039.9578
37	0.2343	19.1426	4.2681	81.7022	0.0522	0.0122	261.8399	13.6784	1117.5562
38	0.2253	19.3679	4.4388	85.9703	0.0516	0.0116	270.1754	13.9497	1199.2584
39	0.2166	19.5845	4.6164	90.4091	0.0511	0.0111	278.4070	14.2157	1285.2287
40	0.2083	19.7928	4.8010	95.0255	0.0505	0.0105	286.5303	14.4765	1375.6379
41	0.2003	19.9931	4.9931	99.8265	0.0500	0.0100	294.5414	14.7322	1470.6634
42	0.1926	20.1856	5.1928	104.8196	0.0495	0.0095	302.4370	14.9828	1570.4899
43	0.1852	20.3708	5.4005	110.0124	0.0491	0.0091	310.2141	15.2284	1675.3095
44	0.1780	20.5488	5.6165	115.4129	0.0487	0.0087	317.8700	15.4690	1785.3219
45	0.1712	20.7200	5.8412	121.0294	0.0483	0.0083	325.4028	15.7047	1900.7348
46	0.1646	20.8847	6.0748	126.8706	0.0479	0.0079	332.8104	15.9356	2021.7642
47	0.1583	21.0429	6.3178	132.9454	0.0475	0.0075	340.0914	16.1618	2148.6348
48	0.1522	21.1951	6.5705	139.2632	0.0472	0.0072	347.2446	16.3832	2281.5802
49	0.1463	21.3415	6.8333	145.8337	0.0469	0.0069	354.2689	16.6000	2420.8434
50	0.1407	21.4822	7.1067	152.6671	0.0466	0.0066	361.1638	16.8122	2566.6771
51	0.1353	21.6175	7.3910	159.7738	0.0463	0.0063	367.9289	17.0200	2719.3442
52	0.1301	21.7476	7.6866	167.1647	0.0460	0.0060	374.5638	17.2232	2879.1179
53	0.1251	21.8727	7.9941	174.8513	0.0457	0.0057	381.0686	17.4221	3046.2827
54	0.1203	21.9930	8.3138	182.8454	0.0455	0.0055	387.4436	17.6167	3221.1340
55	0.1157	22.1086	8.6464	191.1592	0.0452	0.0052	393.6890	17.8070	3403.9793
56	0.1112	22.2198	8.9922	199.8055	0.0450	0.0050	399.8054	17.9932	3595.1385
57	0.1069	22.3267	9.3519	208.7978	0.0448	0.0048	405.7935	18.1752	3794.9440
58	0.1028	22.4296	9.7260	218.1497	0.0446	0.0046	411.6540	18.3532	4003.7418
59	0.0989	22.5284	10.1150	227.8757	0.0444	0.0044	417.3881	18.5272	4221.8915
60	0.0951	22.6235	10.5196	237.9907	0.0442	0.0042	422.9966	18.6972	4449.7671
65	0.0781	23.0467	12.7987	294.9684	0.0434	0.0034	449.2014	19.4909	5749.2095
70	0.0642	23.3945	15.5716	364.2905	0.0427	0.0027	472.4789	20.1961	7357.2615
75	0.0528	23.6804	18.9453	448.6314	0.0422	0.0022	493.0408	20.8206	9340.7842
80	0.0434	23.9154	23.0498	551.2450	0.0418	0.0018	511.1161	21.3718	11781.1244

Financial Factors, Interest Rate, i = 4%

N	(P/F)	(P/A)	(F/P)	(F/A)	(A/P)	(A/F)	(P/G)	(A/G)	(F/G)
85	0.0357	24.1085	28.0436	676.0901	0.0415	0.0015	526.9384	21.8569	14777.2531
90	0.0293	24.2673	34.1193	827.9833	0.0412	0.0012	540.7369	22.2826	18449.5833
95	0.0241	24.3978	41.5114	1012.7846	0.0410	0.0010	552.7307	22.6550	22944.6162
100	0.0198	24.5050	50.5049	1237.6237	0.0408	0.0008	563.1249	22.9800	28440.5926

Financial Factors, Interest Rate, i = 5%

N	(P/F)	(P/A)	(F/P)	(F/A)	(A/P)	(A/F)	(P/G)	(A/G)	(F/G)
1	0.9524	0.9524	1.0500	1.0000	1.0500	1.0000	0.0000	0.0000	0.0000
2	0.9070	1.8594	1.1025	2.0500	0.5378	0.4878	0.9070	0.4878	1.0000
3	0.8638	2.7232	1.1576	3.1525	0.3672	0.3172	2.6347	0.9675	3.0500
4	0.8227	3.5460	1.2155	4.3101	0.2820	0.2320	5.1028	1.4391	6.2025
5	0.7835	4.3295	1.2763	5.5256	0.2310	0.1810	8.2369	1.9025	10.5126
6	0.7462	5.0757	1.3401	6.8019	0.1970	0.1470	11.9680	2.3579	16.0383
7	0.7107	5.7864	1.4071	8.1420	0.1728	0.1228	16.2321	2.8052	22.8402
8	0.6768	6.4632	1.4775	9.5491	0.1547	0.1047	20.9700	3.2445	30.9822
9	0.6446	7.1078	1.5513	11.0266	0.1407	0.0907	26.1268	3.6758	40.5313
10	0.6139	7.7217	1.6289	12.5779	0.1295	0.0795	31.6520	4.0991	51.5579
11	0.5847	8.3064	1.7103	14.2068	0.1204	0.0704	37.4988	4.5144	64.1357
12	0.5568	8.8633	1.7959	15.9171	0.1128	0.0628	43.6241	4.9219	78.3425
13	0.5303	9.3936	1.8856	17.7130	0.1065	0.0565	49.9879	5.3215	94.2597
14	0.5051	9.8986	1.9799	19.5986	0.1010	0.0510	56.5538	5.7133	111.9726
15	0.4810	10.3797	2.0789	21.5786	0.0963	0.0463	63.2880	6.0973	131.5713
16	0.4581	10.8378	2.1829	23.6575	0.0923	0.0423	70.1597	6.4736	153.1498
17	0.4363	11.2741	2.2920	25.8404	0.0887	0.0387	77.1405	6.8423	176.8073
18	0.4155	11.6896	2.4066	28.1324	0.0855	0.0355	84.2043	7.2034	202.6477
19	0.3957	12.0853	2.5270	30.5390	0.0827	0.0327	91.3275	7.5569	230.7801
20	0.3769	12.4622	2.6533	33.0660	0.0802	0.0302	98.4884	7.9030	261.3191
21	0.3589	12.8212	2.7860	35.7193	0.0780	0.0280	105.6673	8.2416	294.3850
22	0.3418	13.1630	2.9253	38.5052	0.0760	0.0260	112.8461	8.5730	330.1043
23	0.3256	13.4886	3.0715	41.4305	0.0741	0.0241	120.0087	8.8971	368.6095
24	0.3101	13.7986	3.2251	44.5020	0.0725	0.0225	127.1402	9.2140	410.0400
25	0.2953	14.0939	3.3864	47.7271	0.0710	0.0210	134.2275	9.5238	454.5420
26	0.2812	14.3752	3.5557	51.1135	0.0696	0.0196	141.2585	9.8266	502.2691
27	0.2678	14.6430	3.7335	54.6691	0.0683	0.0183	148.2226	10.1224	553.3825
28	0.2551	14.8981	3.9201	58.4026	0.0671	0.0171	155.1101	10.4114	608.0517
29	0.2429	15.1411	4.1161	62.3227	0.0660	0.0160	161.9126	10.6936	666.4542
30	0.2314	15.3725	4.3219	66.4388	0.0651	0.0151	168.6226	10.9691	728.7770
31	0.2204	15.5928	4.5380	70.7608	0.0641	0.0141	175.2333	11.2381	795.2158
32	0.2099	15.8027	4.7649	75.2988	0.0633	0.0133	181.7392	11.5005	865.9766
33	0.1999	16.0025	5.0032	80.0638	0.0625	0.0125	188.1351	11.7566	941.2754
34	0.1904	16.1929	5.2533	85.0670	0.0618	0.0118	194.4168	12.0063	1021.3392
35	0.1813	16.3742	5.5160	90.3203	0.0611	0.0111	200.5807	12.2498	1106.4061
36	0.1727	16.5469	5.7918	95.8363	0.0604	0.0104	206.6237	12.4872	1196.7265
37	0.1644	16.7113	6.0814	101.6281	0.0598	0.0098	212.5434	12.7186	1292.5628
38	0.1566	16.8679	6.3855	107.7095	0.0593	0.0093	218.3378	12.9440	1394.1909
39	0.1491	17.0170	6.7048	114.0950	0.0588	0.0088	224.0054	13.1636	1501.9005
40	0.1420	17.1591	7.0400	120.7998	0.0583	0.0083	229.5452	13.3775	1615.9955
41	0.1353	17.2944	7.3920	127.8398	0.0578	0.0078	234.9564	13.5857	1736.7953
42	0.1288	17.4232	7.7616	135.2318	0.0574	0.0074	240.2389	13.7884	1864.6350
43	0.1227	17.5459	8.1497	142.9933	0.0570	0.0070	245.3925	13.9857	1999.8668
44	0.1169	17.6628	8.5572	151.1430	0.0566	0.0066	250.4175	14.1777	2142.8601

Financial Factors, Interest Rate, i = 5%

N	(P/F)	(P/A)	(F/P)	(F/A)	(A/P)	(A/F)	(P/G)	(A/G)	(F/G)
45	0.1113	17.7741	8.9850	159.7002	0.0563	0.0063	255.3145	14.3644	2294.0031
46	0.1060	17.8801	9.4343	168.6852	0.0559	0.0059	260.0844	14.5461	2453.7033
47	0.1009	17.9810	9.9060	178.1194	0.0556	0.0056	264.7281	14.7226	2622.3884
48	0.0961	18.0772	10.4013	188.0254	0.0553	0.0053	269.2467	14.8943	2800.5079
49	0.0916	18.1687	10.9213	198.4267	0.0550	0.0050	273.6418	15.0611	2988.5333
50	0.0872	18.2559	11.4674	209.3480	0.0548	0.0048	277.9148	15.2233	3186.9599
51	0.0831	18.3390	12.0408	220.8154	0.0545	0.0045	282.0673	15.3808	3396.3079
52	0.0791	18.4181	12.6428	232.8562	0.0543	0.0043	286.1013	15.5337	3617.1233
53	0.0753	18.4934	13.2749	245.4990	0.0541	0.0041	290.0184	15.6823	3849.9795
54	0.0717	18.5651	13.9387	258.7739	0.0539	0.0039	293.8208	15.8265	4095.4784
55	0.0683	18.6335	14.6356	272.7126	0.0537	0.0037	297.5104	15.9664	4354.2524
56	0.0651	18.6985	15.3674	287.3482	0.0535	0.0035	301.0894	16.1023	4626.9650
57	0.0620	18.7605	16.1358	302.7157	0.0533	0.0033	304.5599	16.2341	4914.3132
58	0.0590	18.8195	16.9426	318.8514	0.0531	0.0031	307.9243	16.3619	5217.0289
59	0.0562	18.8758	17.7897	335.7940	0.0530	0.0030	311.1846	16.4859	5535.8803
60	0.0535	18.9293	18.6792	353.5837	0.0528	0.0028	314.3432	16.6062	5871.6744
65	0.0419	19.1611	23.8399	456.7980	0.0522	0.0022	328.6910	17.1541	7835.9602
70	0.0329	19.3427	30.4264	588.5285	0.0517	0.0017	340.8409	17.6212	10370.5702
75	0.0258	19.4850	38.8327	756.6537	0.0513	0.0013	351.0721	18.0176	13633.0744
80	0.0202	19.5965	49.5614	971.2288	0.0510	0.0010	359.6460	18.3526	17824.5764
85	0.0158	19.6838	63.2544	1245.0871	0.0508	0.0008	366.8007	18.6346	23201.7414
90	0.0124	19.7523	80.7304	1594.6073	0.0506	0.0006	372.7488	18.8712	30092.1460
95	0.0097	19.8059	103.0347	2040.6935	0.0505	0.0005	377.6774	19.0689	38913.8706
100	0.0076	19.8479	131.5013	2610.0252	0.0504	0.0004	381.7492	19.2337	50200.5031

Financial Factors, Interest Rate, i = 6%

N	(P/F)	(P/A)	(F/P)	(F/A)	(A/P)	(A/F)	(P/G)	(A/G)	(F/G)
1	0.9434	0.9434	1.0600	1.0000	1.0600	1.0000	0.0000	0.0000	0.0000
2	0.8900	1.8334	1.1236	2.0600	0.5454	0.4854	0.8900	0.4854	1.0000
3	0.8396	2.6730	1.1910	3.1836	0.3741	0.3141	2.5692	0.9612	3.0600
4	0.7921	3.4651	1.2625	4.3746	0.2886	0.2286	4.9455	1.4272	6.2436
5	0.7473	4.2124	1.3382	5.6371	0.2374	0.1774	7.9345	1.8836	10.6182
6	0.7050	4.9173	1.4185	6.9753	0.2034	0.1434	11.4594	2.3304	16.2553
7	0.6651	5.5824	1.5036	8.3938	0.1791	0.1191	15.4497	2.7676	23.2306
8	0.6274	6.2098	1.5938	9.8975	0.1610	0.1010	19.8416	3.1952	31.6245
9	0.5919	6.8017	1.6895	11.4913	0.1470	0.0870	24.5768	3.6133	41.5219
10	0.5584	7.3601	1.7908	13.1808	0.1359	0.0759	29.6023	4.0220	53.0132
11	0.5268	7.8869	1.8983	14.9716	0.1268	0.0668	34.8702	4.4213	66.1940
12	0.4970	8.3838	2.0122	16.8699	0.1193	0.0593	40.3369	4.8113	81.1657
13	0.4688	8.8527	2.1329	18.8821	0.1130	0.0530	45.9629	5.1920	98.0356
14	0.4423	9.2950	2.2609	21.0151	0.1076	0.0476	51.7128	5.5635	116.9178
15	0.4173	9.7122	2.3966	23.2760	0.1030	0.0430	57.5546	5.9260	137.9328
16	0.3936	10.1059	2.5404	25.6725	0.0990	0.0390	63.4592	6.2794	161.2088
17	0.3714	10.4773	2.6928	28.2129	0.0954	0.0354	69.4011	6.6240	186.8813
18	0.3503	10.8276	2.8543	30.9057	0.0924	0.0324	75.3569	6.9597	215.0942
19	0.3305	11.1581	3.0256	33.7600	0.0896	0.0296	81.3062	7.2867	245.9999
20	0.3118	11.4699	3.2071	36.7856	0.0872	0.0272	87.2304	7.6051	279.7599
21	0.2942	11.7641	3.3996	39.9927	0.0850	0.0250	93.1136	7.9151	316.5454
22	0.2775	12.0416	3.6035	43.3923	0.0830	0.0230	98.9412	8.2166	356.5382
23	0.2618	12.3034	3.8197	46.9958	0.0813	0.0213	104.7007	8.5099	399.9305

Financial Factors, Interest Rate, i = 6%

N	(P/F)	(P/A)	(F/P)	(F/A)	(A/P)	(A/F)	(P/G)	(A/G)	(F/G)
24	0.2470	12.5504	4.0489	50.8156	0.0797	0.0197	110.3812	8.7951	446.9263
25	0.2330	12.7834	4.2919	54.8645	0.0782	0.0182	115.9732	9.0722	497.7419
26	0.2198	13.0032	4.5494	59.1564	0.0769	0.0169	121.4684	9.3414	552.6064
27	0.2074	13.2105	4.8223	63.7058	0.0757	0.0157	126.8600	9.6029	611.7628
28	0.1956	13.4062	5.1117	68.5281	0.0746	0.0146	132.1420	9.8568	675.4685
29	0.1846	13.5907	5.4184	73.6398	0.0736	0.0136	137.3096	10.1032	743.9966
30	0.1741	13.7648	5.7435	79.0582	0.0726	0.0126	142.3588	10.3422	817.6364
31	0.1643	13.9291	6.0881	84.8017	0.0718	0.0118	147.2864	10.5740	896.6946
32	0.1550	14.0840	6.4534	90.8898	0.0710	0.0110	152.0901	10.7988	981.4963
33	0.1462	14.2302	6.8406	97.3432	0.0703	0.0103	156.7681	11.0166	1072.3861
34	0.1379	14.3681	7.2510	104.1838	0.0696	0.0096	161.3192	11.2276	1169.7292
35	0.1301	14.4982	7.6861	111.4348	0.0690	0.0090	165.7427	11.4319	1273.9130
36	0.1227	14.6210	8.1473	119.1209	0.0684	0.0084	170.0387	11.6298	1385.3478
37	0.1158	14.7368	8.6361	127.2681	0.0679	0.0079	174.2072	11.8213	1504.4686
38	0.1092	14.8460	9.1543	135.9042	0.0674	0.0074	178.2490	12.0065	1631.7368
39	0.1031	14.9491	9.7035	145.0585	0.0669	0.0069	182.1652	12.1857	1767.6410
40	0.0972	15.0463	10.2857	154.7620	0.0665	0.0065	185.9568	12.3590	1912.6994
41	0.0917	15.1380	10.9029	165.0477	0.0661	0.0061	189.6256	12.5264	2067.4614
42	0.0865	15.2245	11.5570	175.9505	0.0657	0.0057	193.1732	12.6883	2232.5091
43	0.0816	15.3062	12.2505	187.5076	0.0653	0.0053	196.6017	12.8446	2408.4596
44	0.0770	15.3832	12.9855	199.7580	0.0650	0.0050	199.9130	12.9956	2595.9672
45	0.0727	15.4558	13.7646	212.7435	0.0647	0.0047	203.1096	13.1413	2795.7252
46	0.0685	15.5244	14.5905	226.5081	0.0644	0.0044	206.1938	13.2819	3008.4687
47	0.0647	15.5890	15.4659	241.0986	0.0641	0.0041	209.1681	13.4177	3234.9769
48	0.0610	15.6500	16.3939	256.5645	0.0639	0.0039	212.0351	13.5485	3476.0755
49	0.0575	15.7076	17.3775	272.9584	0.0637	0.0037	214.7972	13.6748	3732.6400
50	0.0543	15.7619	18.4202	290.3359	0.0634	0.0034	217.4574	13.7964	4005.5984
51	0.0512	15.8131	19.5254	308.7561	0.0632	0.0032	220.0181	13.9137	4295.9343
52	0.0483	15.8614	20.6969	328.2814	0.0630	0.0030	222.4823	14.0267	4604.6904
53	0.0456	15.9070	21.9387	348.9783	0.0629	0.0029	224.8525	14.1355	4932.9718
54	0.0430	15.9500	23.2550	370.9170	0.0627	0.0027	227.1316	14.2402	5281.9501
55	0.0406	15.9905	24.6503	394.1720	0.0625	0.0025	229.3222	14.3411	5652.8671
56	0.0383	16.0288	26.1293	418.8223	0.0624	0.0024	231.4272	14.4382	6047.0391
57	0.0361	16.0649	27.6971	444.9517	0.0622	0.0022	233.4490	14.5316	6465.8615
58	0.0341	16.0990	29.3589	472.6488	0.0621	0.0021	235.3905	14.6215	6910.8132
59	0.0321	16.1311	31.1205	502.0077	0.0620	0.0020	237.2542	14.7079	7383.4620
60	0.0303	16.1614	32.9877	533.1282	0.0619	0.0019	239.0428	14.7909	7885.4697
65	0.0227	16.2891	44.1450	719.0829	0.0614	0.0014	246.9450	15.1601	10901.3810
70	0.0169	16.3845	59.0759	967.9322	0.0610	0.0010	253.3271	15.4613	14965.5362
75	0.0126	16.4558	79.0569	1300.9487	0.0608	0.0008	258.4527	15.7058	20432.4780
80	0.0095	16.5091	105.7960	1746.5999	0.0606	0.0006	262.5493	15.9033	27776.6649
85	0.0071	16.5489	141.5789	2342.9817	0.0604	0.0004	265.8096	16.0620	37633.0290
90	0.0053	16.5787	189.4645	3141.0752	0.0603	0.0003	268.3946	16.1891	50851.2531
95	0.0039	16.6009	253.5463	4209.1042	0.0602	0.0002	270.4375	16.2905	68568.4042
100	0.0029	16.6175	339.3021	5638.3681	0.0602	0.0002	272.0471	16.3711	92306.1343

Financial Factors, Interest Rate, i = 7%

N	(P/F)	(P/A)	(F/P)	(F/A)	(A/P)	(A/F)	(P/G)	(A/G)	(F/G)
1	0.9346	0.9346	1.0700	1.0000	1.0700	1.0000	0.0000	0.0000	0.0000
2	0.8734	1.8080	1.1449	2.0700	0.5531	0.4831	0.8734	0.4831	1.0000
3	0.8163	2.6243	1.2250	3.2149	0.3811	0.3111	2.5060	0.9549	3.0700

Financial Factors, Interest Rate, i = 7%

N	(P/F)	(P/A)	(F/P)	(F/A)	(A/P)	(A/F)	(P/G)	(A/G)	(F/G)
4	0.7629	3.3872	1.3108	4.4399	0.2952	0.2252	4.7947	1.4155	6.2849
5	0.7130	4.1002	1.4026	5.7507	0.2439	0.1739	7.6467	1.8650	10.7248
6	0.6663	4.7665	1.5007	7.1533	0.2098	0.1398	10.9784	2.3032	16.4756
7	0.6227	5.3893	1.6058	8.6540	0.1856	0.1156	14.7149	2.7304	23.6289
8	0.5820	5.9713	1.7182	10.2598	0.1675	0.0975	18.7889	3.1465	32.2829
9	0.5439	6.5152	1.8385	11.9780	0.1535	0.0835	23.1404	3.5517	42.5427
10	0.5083	7.0236	1.9672	13.8164	0.1424	0.0724	27.7156	3.9461	54.5207
11	0.4751	7.4987	2.1049	15.7836	0.1334	0.0634	32.4665	4.3296	68.3371
12	0.4440	7.9427	2.2522	17.8885	0.1259	0.0559	37.3506	4.7025	84.1207
13	0.4150	8.3577	2.4098	20.1406	0.1197	0.0497	42.3302	5.0648	102.0092
14	0.3878	8.7455	2.5785	22.5505	0.1143	0.0443	47.3718	5.4167	122.1498
15	0.3624	9.1079	2.7590	25.1290	0.1098	0.0398	52.4461	5.7583	144.7003
16	0.3387	9.4466	2.9522	27.8881	0.1059	0.0359	57.5271	6.0897	169.8293
17	0.3166	9.7632	3.1588	30.8402	0.1024	0.0324	62.5923	6.4110	197.7174
18	0.2959	10.0591	3.3799	33.9990	0.0994	0.0294	67.6219	6.7225	228.5576
19	0.2765	10.3356	3.6165	37.3790	0.0968	0.0268	72.5991	7.0242	262.5566
20	0.2584	10.5940	3.8697	40.9955	0.0944	0.0244	77.5091	7.3163	299.9356
21	0.2415	10.8355	4.1406	44.8652	0.0923	0.0223	82.3393	7.5990	340.9311
22	0.2257	11.0612	4.4304	49.0057	0.0904	0.0204	87.0793	7.8725	385.7963
23	0.2109	11.2722	4.7405	53.4361	0.0887	0.0187	91.7201	8.1369	434.8020
24	0.1971	11.4693	5.0724	58.1767	0.0872	0.0172	96.2545	8.3923	488.2382
25	0.1842	11.6536	5.4274	63.2490	0.0858	0.0158	100.6765	8.6391	546.4148
26	0.1722	11.8258	5.8074	68.6765	0.0846	0.0146	104.9814	8.8773	609.6639
27	0.1609	11.9867	6.2139	74.4838	0.0834	0.0134	109.1656	9.1072	678.3403
28	0.1504	12.1371	6.6488	80.6977	0.0824	0.0124	113.2264	9.3289	752.8242
29	0.1406	12.2777	7.1143	87.3465	0.0814	0.0114	117.1622	9.5427	833.5218
30	0.1314	12.4090	7.6123	94.4608	0.0806	0.0106	120.9718	9.7487	920.8684
31	0.1228	12.5318	8.1451	102.0730	0.0798	0.0098	124.6550	9.9471	1015.3292
32	0.1147	12.6466	8.7153	110.2182	0.0791	0.0091	128.2120	10.1381	1117.4022
33	0.1072	12.7538	9.3253	118.9334	0.0784	0.0084	131.6435	10.3219	1227.6204
34	0.1002	12.8540	9.9781	128.2588	0.0778	0.0078	134.9507	10.4987	1346.5538
35	0.0937	12.9477	10.6766	138.2369	0.0772	0.0072	138.1353	10.6687	1474.8125
36	0.0875	13.0352	11.4239	148.9135	0.0767	0.0067	141.1990	10.8321	1613.0494
37	0.0818	13.1170	12.2236	160.3374	0.0762	0.0062	144.1441	10.9891	1761.9629
38	0.0765	13.1935	13.0793	172.5610	0.0758	0.0058	146.9730	11.1398	1922.3003
39	0.0715	13.2649	13.9948	185.6403	0.0754	0.0054	149.6883	11.2845	2094.8613
40	0.0668	13.3317	14.9745	199.6351	0.0750	0.0050	152.2928	11.4233	2280.5016
41	0.0624	13.3941	16.0227	214.6096	0.0747	0.0047	154.7892	11.5565	2480.1367
42	0.0583	13.4524	17.1443	230.6322	0.0743	0.0043	157.1807	11.6842	2694.7463
43	0.0545	13.5070	18.3444	247.7765	0.0740	0.0040	159.4702	11.8065	2925.3785
44	0.0509	13.5579	19.6285	266.1209	0.0738	0.0038	161.6609	11.9237	3173.1550
45	0.0476	13.6055	21.0025	285.7493	0.0735	0.0035	163.7559	12.0360	3439.2759
46	0.0445	13.6500	22.4726	306.7518	0.0733	0.0033	165.7584	12.1435	3725.0252
47	0.0416	13.6916	24.0457	329.2244	0.0730	0.0030	167.6714	12.2463	4031.7769
48	0.0389	13.7305	25.7289	353.2701	0.0728	0.0028	169.4981	12.3447	4361.0013
49	0.0363	13.7668	27.5299	378.9990	0.0726	0.0026	171.2417	12.4387	4714.2714
50	0.0339	13.8007	29.4570	406.5289	0.0725	0.0025	172.9051	12.5287	5093.2704
51	0.0317	13.8325	31.5190	435.9860	0.0723	0.0023	174.4915	12.6146	5499.7994
52	0.0297	13.8621	33.7253	467.5050	0.0721	0.0021	176.0037	12.6967	5935.7853
53	0.0277	13.8898	36.0861	501.2303	0.0720	0.0020	177.4447	12.7751	6403.2903
54	0.0259	13.9157	38.6122	537.3164	0.0719	0.0019	178.8173	12.8500	6904.5206
55	0.0242	13.9399	41.3150	575.9286	0.0717	0.0017	180.1243	12.9215	7441.8370
56	0.0226	13.9626	44.2071	617.2436	0.0716	0.0016	181.3685	12.9896	8017.7656
57	0.0211	13.9837	47.3015	661.4506	0.0715	0.0015	182.5524	13.0547	8635.0092

Financial Factors, Interest Rate, i = 7%

N	(P/F)	(P/A)	(F/P)	(F/A)	(A/P)	(A/F)	(P/G)	(A/G)	(F/G)
58	0.0198	14.0035	50.6127	708.7522	0.0714	0.0014	183.6786	13.1167	9296.4599
59	0.0185	14.0219	54.1555	759.3648	0.0713	0.0013	184.7496	13.1758	10005.2121
60	0.0173	14.0392	57.9464	813.5204	0.0712	0.0012	185.7677	13.2321	10764.5769
65	0.0123	14.1099	81.2729	1146.7552	0.0709	0.0009	190.1452	13.4760	15453.6452
70	0.0088	14.1604	113.9894	1614.1342	0.0706	0.0006	193.5185	13.6662	22059.0596
75	0.0063	14.1964	159.8760	2269.6574	0.0704	0.0004	196.1035	13.8136	31352.2488
80	0.0045	14.2220	224.2344	3189.0627	0.0703	0.0003	198.0748	13.9273	44415.1811
85	0.0032	14.2403	314.5003	4478.5761	0.0702	0.0002	199.5717	14.0146	62765.3731
90	0.0023	14.2533	441.1030	6287.1854	0.0702	0.0002	200.7042	14.0812	88531.2204
95	0.0016	14.2626	618.6697	8823.8535	0.0701	0.0001	201.5581	14.1319	124697.9077
100	0.0012	14.2693	867.7163	12381.6618	0.0701	0.0001	202.2001	14.1703	175452.3113

Financial Factors, Interest Rate, i = 8%

N	(P/F)	(P/A)	(F/P)	(F/A)	(A/P)	(A/F)	(P/G)	(A/G)	(F/G)
1	0.9259	0.9259	1.0800	1.0000	1.0800	1.0000	0.0000	0.0000	0.0000
2	0.8573	1.7833	1.1664	2.0800	0.5608	0.4808	0.8573	0.4808	1.0000
3	0.7938	2.5771	1.2597	3.2464	0.3880	0.3080	2.4450	0.9487	3.0800
4	0.7350	3.3121	1.3605	4.5061	0.3019	0.2219	4.6501	1.4040	6.3264
5	0.6806	3.9927	1.4693	5.8666	0.2505	0.1705	7.3724	1.8465	10.8325
6	0.6302	4.6229	1.5869	7.3359	0.2163	0.1363	10.5233	2.2763	16.6991
7	0.5835	5.2064	1.7138	8.9228	0.1921	0.1121	14.0242	2.6937	24.0350
8	0.5403	5.7466	1.8509	10.6366	0.1740	0.0940	17.8061	3.0985	32.9578
9	0.5002	6.2469	1.9990	12.4876	0.1601	0.0801	21.8081	3.4910	43.5945
10	0.4632	6.7101	2.1589	14.4866	0.1490	0.0690	25.9768	3.8713	56.0820
11	0.4289	7.1390	2.3316	16.6455	0.1401	0.0601	30.2657	4.2395	70.5686
12	0.3971	7.5361	2.5182	18.9771	0.1327	0.0527	34.6339	4.5957	87.2141
13	0.3677	7.9038	2.7196	21.4953	0.1265	0.0465	39.0463	4.9402	106.1912
14	0.3405	8.2442	2.9372	24.2149	0.1213	0.0413	43.4723	5.2731	127.6865
15	0.3152	8.5595	3.1722	27.1521	0.1168	0.0368	47.8857	5.5945	151.9014
16	0.2919	8.8514	3.4259	30.3243	0.1130	0.0330	52.2640	5.9046	179.0535
17	0.2703	9.1216	3.7000	33.7502	0.1096	0.0296	56.5883	6.2037	209.3778
18	0.2502	9.3719	3.9960	37.4502	0.1067	0.0267	60.8426	6.4920	243.1280
19	0.2317	9.6036	4.3157	41.4463	0.1041	0.0241	65.0134	6.7697	280.5783
20	0.2145	9.8181	4.6610	45.7620	0.1019	0.0219	69.0898	7.0369	322.0246
21	0.1987	10.0168	5.0338	50.4229	0.0998	0.0198	73.0629	7.2940	367.7865
22	0.1839	10.2007	5.4365	55.4568	0.0980	0.0180	76.9257	7.5412	418.2094
23	0.1703	10.3711	5.8715	60.8933	0.0964	0.0164	80.6726	7.7786	473.6662
24	0.1577	10.5288	6.3412	66.7648	0.0950	0.0150	84.2997	8.0066	534.5595
25	0.1460	10.6748	6.8485	73.1059	0.0937	0.0137	87.8041	8.2254	601.3242
26	0.1352	10.8100	7.3964	79.9544	0.0925	0.0125	91.1842	8.4352	674.4302
27	0.1252	10.9352	7.9881	87.3508	0.0914	0.0114	94.4390	8.6363	754.3846
28	0.1159	11.0511	8.6271	95.3388	0.0905	0.0105	97.5687	8.8289	841.7354
29	0.1073	11.1584	9.3173	103.9659	0.0896	0.0096	100.5738	9.0133	937.0742
30	0.0994	11.2578	10.0627	113.2832	0.0888	0.0088	103.4558	9.1897	1041.0401
31	0.0920	11.3498	10.8677	123.3459	0.0881	0.0081	106.2163	9.3584	1154.3234
32	0.0852	11.4350	11.7371	134.2135	0.0875	0.0075	108.8575	9.5197	1277.6692
33	0.0789	11.5139	12.6760	145.9506	0.0869	0.0069	111.3819	9.6737	1411.8828
34	0.0730	11.5869	13.6901	158.6267	0.0863	0.0063	113.7924	9.8208	1557.8334
35	0.0676	11.6546	14.7853	172.3168	0.0858	0.0058	116.0920	9.9611	1716.4600
36	0.0626	11.7172	15.9682	187.1021	0.0853	0.0053	118.2839	10.0949	1888.7768
37	0.0580	11.7752	17.2456	203.0703	0.0849	0.0049	120.3713	10.2225	2075.8790

Financial Factors, Interest Rate, i = 8%

N	(P/F)	(P/A)	(F/P)	(F/A)	(A/P)	(A/F)	(P/G)	(A/G)	(F/G)
38	0.0537	11.8289	18.6253	220.3159	0.0845	0.0045	122.3579	10.3440	2278.9493
39	0.0497	11.8786	20.1153	238.9412	0.0842	0.0042	124.2470	10.4597	2499.2653
40	0.0460	11.9246	21.7245	259.0565	0.0839	0.0039	126.0422	10.5699	2738.2065
41	0.0426	11.9672	23.4625	280.7810	0.0836	0.0036	127.7470	10.6747	2997.2630
42	0.0395	12.0067	25.3395	304.2435	0.0833	0.0033	129.3651	10.7744	3278.0440
43	0.0365	12.0432	27.3666	329.5830	0.0830	0.0030	130.8998	10.8692	3582.2876
44	0.0338	12.0771	29.5560	356.9496	0.0828	0.0028	132.3547	10.9592	3911.8706
45	0.0313	12.1084	31.9204	386.5056	0.0826	0.0026	133.7331	11.0447	4268.8202
46	0.0290	12.1374	34.4741	418.4261	0.0824	0.0024	135.0384	11.1258	4655.3258
47	0.0269	12.1643	37.2320	452.9002	0.0822	0.0022	136.2739	11.2028	5073.7519
48	0.0249	12.1891	40.2106	490.1322	0.0820	0.0020	137.4428	11.2758	5526.6521
49	0.0230	12.2122	43.4274	530.3427	0.0819	0.0019	138.5480	11.3451	6016.7842
50	0.0213	12.2335	46.9016	573.7702	0.0817	0.0017	139.5928	11.4107	6547.1270
51	0.0197	12.2532	50.6537	620.6718	0.0816	0.0016	140.5799	11.4729	7120.8971
52	0.0183	12.2715	54.7060	671.3255	0.0815	0.0015	141.5121	11.5318	7741.5689
53	0.0169	12.2884	59.0825	726.0316	0.0814	0.0014	142.3923	11.5875	8412.8944
54	0.0157	12.3041	63.8091	785.1141	0.0813	0.0013	143.2229	11.6403	9138.9259
55	0.0145	12.3186	68.9139	848.9232	0.0812	0.0012	144.0065	11.6902	9924.0400
56	0.0134	12.3321	74.4270	917.8371	0.0811	0.0011	144.7454	11.7373	10772.9632
57	0.0124	12.3445	80.3811	992.2640	0.0810	0.0010	145.4421	11.7819	11690.8003
58	0.0115	12.3560	86.8116	1072.6451	0.0809	0.0009	146.0987	11.8241	12683.0643
59	0.0107	12.3667	93.7565	1159.4568	0.0809	0.0009	146.7173	11.8639	13755.7094
60	0.0099	12.3766	101.2571	1253.2133	0.0808	0.0008	147.3000	11.9015	14915.1662
65	0.0067	12.4160	148.7798	1847.2481	0.0805	0.0005	149.7387	12.0602	22278.1010
70	0.0046	12.4428	218.6064	2720.0801	0.0804	0.0004	151.5326	12.1783	33126.0009
75	0.0031	12.4611	321.2045	4002.5566	0.0802	0.0002	152.8448	12.2658	49094.4578
80	0.0021	12.4735	471.9548	5886.9354	0.0802	0.0002	153.8001	12.3301	72586.6929
85	0.0014	12.4820	693.4565	8655.7061	0.0801	0.0001	154.4925	12.3772	107133.8264
90	0.0010	12.4877	1018.9151	12723.9386	0.0801	0.0001	154.9925	12.4116	157924.2327
95	0.0007	12.4917	1497.1205	18701.5069	0.0801	0.0001	155.3524	12.4365	232581.3357
100	0.0005	12.4943	2199.7613	27484.5157	0.0800	0.0000	155.6107	12.4545	342306.4463

Financial Factors, Interest Rate, i = 9%

N	(P/F)	(P/A)	(F/P)	(F/A)	(A/P)	(A/F)	(P/G)	(A/G)	(F/G)
1	0.9174	0.9174	1.0900	1.0000	1.0900	1.0000	0.0000	0.0000	0.0000
2	0.8417	1.7591	1.1881	2.0900	0.5685	0.4785	0.8417	0.4785	1.0000
3	0.7722	2.5313	1.2950	3.2781	0.3951	0.3051	2.3860	0.9426	3.0900
4	0.7084	3.2397	1.4116	4.5731	0.3087	0.2187	4.5113	1.3925	6.3681
5	0.6499	3.8897	1.5386	5.9847	0.2571	0.1671	7.1110	1.8282	10.9412
6	0.5963	4.4859	1.6771	7.5233	0.2229	0.1329	10.0924	2.2498	16.9259
7	0.5470	5.0330	1.8280	9.2004	0.1987	0.1087	13.3746	2.6574	24.4493
8	0.5019	5.5348	1.9926	11.0285	0.1807	0.0907	16.8877	3.0512	33.6497
9	0.4604	5.9952	2.1719	13.0210	0.1668	0.0768	20.5711	3.4312	44.6782
10	0.4224	6.4177	2.3674	15.1929	0.1558	0.0658	24.3728	3.7978	57.6992
11	0.3875	6.8052	2.5804	17.5603	0.1469	0.0569	28.2481	4.1510	72.8921
12	0.3555	7.1607	2.8127	20.1407	0.1397	0.0497	32.1590	4.4910	90.4524
13	0.3262	7.4869	3.0658	22.9534	0.1336	0.0436	36.0731	4.8182	110.5932
14	0.2992	7.7862	3.3417	26.0192	0.1284	0.0384	39.9633	5.1326	133.5465
15	0.2745	8.0607	3.6425	29.3609	0.1241	0.0341	43.8069	5.4346	159.5657
16	0.2519	8.3126	3.9703	33.0034	0.1203	0.0303	47.5849	5.7245	188.9267

Financial Factors, Interest Rate, i = 9%

N	(P/F)	(P/A)	(F/P)	(F/A)	(A/P)	(A/F)	(P/G)	(A/G)	(F/G)
17	0.2311	8.5436	4.3276	36.9737	0.1170	0.0270	51.2821	6.0024	221.9301
18	0.2120	8.7556	4.7171	41.3013	0.1142	0.0242	54.8860	6.2687	258.9038
19	0.1945	8.9501	5.1417	46.0185	0.1117	0.0217	58.3868	6.5236	300.2051
20	0.1784	9.1285	5.6044	51.1601	0.1095	0.0195	61.7770	6.7674	346.2236
21	0.1637	9.2922	6.1088	56.7645	0.1076	0.0176	65.0509	7.0006	397.3837
22	0.1502	9.4424	6.6586	62.8733	0.1059	0.0159	68.2048	7.2232	454.1482
23	0.1378	9.5802	7.2579	69.5319	0.1044	0.0144	71.2359	7.4357	517.0215
24	0.1264	9.7066	7.9111	76.7898	0.1030	0.0130	74.1433	7.6384	586.5535
25	0.1160	9.8226	8.6231	84.7009	0.1018	0.0118	76.9265	7.8316	663.3433
26	0.1064	9.9290	9.3992	93.3240	0.1007	0.0107	79.5863	8.0156	748.0442
27	0.0976	10.0266	10.2451	102.7231	0.0997	0.0097	82.1241	8.1906	841.3682
28	0.0895	10.1161	11.1671	112.9682	0.0989	0.0089	84.5419	8.3571	944.0913
29	0.0822	10.1983	12.1722	124.1354	0.0981	0.0081	86.8422	8.5154	1057.0595
30	0.0754	10.2737	13.2677	136.3075	0.0973	0.0073	89.0280	8.6657	1181.1949
31	0.0691	10.3428	14.4618	149.5752	0.0967	0.0067	91.1024	8.8083	1317.5024
32	0.0634	10.4062	15.7633	164.0370	0.0961	0.0061	93.0690	8.9436	1467.0776
33	0.0582	10.4644	17.1820	179.8003	0.0956	0.0056	94.9314	9.0718	1631.1146
34	0.0534	10.5178	18.7284	196.9823	0.0951	0.0051	96.6935	9.1933	1810.9149
35	0.0490	10.5668	20.4140	215.7108	0.0946	0.0046	98.3590	9.3083	2007.8973
36	0.0449	10.6118	22.2512	236.1247	0.0942	0.0042	99.9319	9.4171	2223.6080
37	0.0412	10.6530	24.2538	258.3759	0.0939	0.0039	101.4162	9.5200	2459.7328
38	0.0378	10.6908	26.4367	282.6298	0.0935	0.0035	102.8158	9.6172	2718.1087
39	0.0347	10.7255	28.8160	309.0665	0.0932	0.0032	104.1345	9.7090	3000.7385
40	0.0318	10.7574	31.4094	337.8824	0.0930	0.0030	105.3762	9.7957	3309.8049
41	0.0292	10.7866	34.2363	369.2919	0.0927	0.0027	106.5445	9.8775	3647.6874
42	0.0268	10.8134	37.3175	403.5281	0.0925	0.0025	107.6432	9.9546	4016.9793
43	0.0246	10.8380	40.6761	440.8457	0.0923	0.0023	108.6758	10.0273	4420.5074
44	0.0226	10.8605	44.3370	481.5218	0.0921	0.0021	109.6456	10.0958	4861.3531
45	0.0207	10.8812	48.3273	525.8587	0.0919	0.0019	110.5561	10.1603	5342.8748
46	0.0190	10.9002	52.6767	574.1860	0.0917	0.0017	111.4103	10.2210	5868.7336
47	0.0174	10.9176	57.4176	626.8628	0.0916	0.0016	112.2115	10.2780	6442.9196
48	0.0160	10.9336	62.5852	684.2804	0.0915	0.0015	112.9625	10.3317	7069.7823
49	0.0147	10.9482	68.2179	746.8656	0.0913	0.0013	113.6661	10.3821	7754.0628
50	0.0134	10.9617	74.3575	815.0836	0.0912	0.0012	114.3251	10.4295	8500.9284
51	0.0123	10.9740	81.0497	889.4411	0.0911	0.0011	114.9420	10.4740	9316.0120
52	0.0113	10.9853	88.3442	970.4908	0.0910	0.0010	115.5193	10.5158	10205.4530
53	0.0104	10.9957	96.2951	1058.8349	0.0909	0.0009	116.0593	10.5549	11175.9438
54	0.0095	11.0053	104.9617	1155.1301	0.0909	0.0009	116.5642	10.5917	12234.7788
55	0.0087	11.0140	114.4083	1260.0918	0.0908	0.0008	117.0362	10.6261	13389.9088
56	0.0080	11.0220	124.7050	1374.5001	0.0907	0.0007	117.4772	10.6584	14650.0006
57	0.0074	11.0294	135.9285	1499.2051	0.0907	0.0007	117.8892	10.6887	16024.5007
58	0.0067	11.0361	148.1620	1635.1335	0.0906	0.0006	118.2739	10.7170	17523.7058
59	0.0062	11.0423	161.4966	1783.2955	0.0906	0.0006	118.6331	10.7435	19158.8393
60	0.0057	11.0480	176.0313	1944.7921	0.0905	0.0005	118.9683	10.7683	20942.1348
65	0.0037	11.0701	270.8460	2998.2885	0.0903	0.0003	120.3344	10.8702	32592.0942
70	0.0024	11.0844	416.7301	4619.2232	0.0902	0.0002	121.2942	10.9427	50546.9242
75	0.0016	11.0938	641.1909	7113.2321	0.0901	0.0001	121.9646	10.9940	78202.5794
80	0.0010	11.0998	986.5517	10950.5741	0.0901	0.0001	122.4306	11.0299	120784.1566
85	0.0007	11.1038	1517.9320	16854.8003	0.0901	0.0001	122.7533	11.0551	186331.1147
90	0.0004	11.1064	2335.5266	25939.1842	0.0900	0.0000	122.9758	11.0726	287213.1583
95	0.0003	11.1080	3593.4971	39916.6350	0.0900	0.0000	123.1287	11.0847	442462.6107
100	0.0002	11.1091	5529.0408	61422.6755	0.0900	0.0000	123.2335	11.0930	681363.0607

Financial Factors, Interest Rate, i =				**10%**					
N	**(P/F)**	**(P/A)**	**(F/P)**	**(F/A)**	**(A/P)**	**(A/F)**	**(P/G)**	**(A/G)**	**(F/G)**
1	0.9091	0.9091	1.100	1.000	1.1000	1.0000	0.0000	0.0000	0.000
2	0.8264	1.7355	1.210	2.100	0.5762	0.4762	0.8264	0.4762	1.000
3	0.7513	2.4869	1.331	3.310	0.4021	0.3021	2.3291	0.9366	3.100
4	0.6830	3.1699	1.464	4.641	0.3155	0.2155	4.3781	1.3812	6.410
5	0.6209	3.7908	1.611	6.105	0.2638	0.1638	6.8618	1.8101	11.051
6	0.5645	4.3553	1.772	7.716	0.2296	0.1296	9.6842	2.2236	17.156
7	0.5132	4.8684	1.949	9.487	0.2054	0.1054	12.7631	2.6216	24.872
8	0.4665	5.3349	2.144	11.436	0.1874	0.0874	16.0287	3.0045	34.359
9	0.4241	5.7590	2.358	13.579	0.1736	0.0736	19.4215	3.3724	45.795
10	0.3855	6.1446	2.594	15.937	0.1627	0.0627	22.8913	3.7255	59.374
11	0.3505	6.4951	2.853	18.531	0.1540	0.0540	26.3963	4.0641	75.312
12	0.3186	6.8137	3.138	21.384	0.1468	0.0468	29.9012	4.3884	93.843
13	0.2897	7.1034	3.452	24.523	0.1408	0.0408	33.3772	4.6988	115.227
14	0.2633	7.3667	3.797	27.975	0.1357	0.0357	36.8005	4.9955	139.750
15	0.2394	7.6061	4.177	31.772	0.1315	0.0315	40.1520	5.2789	167.725
16	0.2176	7.8237	4.595	35.950	0.1278	0.0278	43.4164	5.5493	199.497
17	0.1978	8.0216	5.054	40.545	0.1247	0.0247	46.5819	5.8071	235.447
18	0.1799	8.2014	5.560	45.599	0.1219	0.0219	49.6395	6.0526	275.992
19	0.1635	8.3649	6.116	51.159	0.1195	0.0195	52.5827	6.2861	321.591
20	0.1486	8.5136	6.727	57.275	0.1175	0.0175	55.4069	6.5081	372.750
21	0.1351	8.6487	7.400	64.002	0.1156	0.0156	58.1095	6.7189	430.025
22	0.1228	8.7715	8.140	71.403	0.1140	0.0140	60.6893	6.9189	494.027
23	0.1117	8.8832	8.954	79.543	0.1126	0.0126	63.1462	7.1085	565.430
24	0.1015	8.9847	9.850	88.497	0.1113	0.0113	65.4813	7.2881	644.973
25	0.0923	9.0770	10.835	98.347	0.1102	0.0102	67.6964	7.4580	733.471
26	0.0839	9.1609	11.918	109.182	0.1092	0.0092	69.7940	7.6186	831.818
27	0.0763	9.2372	13.110	121.100	0.1083	0.0083	71.7773	7.7704	940.999
28	0.0693	9.3066	14.421	134.210	0.1075	0.0075	73.6495	7.9137	1062.099
29	0.0630	9.3696	15.863	148.631	0.1067	0.0067	75.4146	8.0489	1196.309
30	0.0573	9.4269	17.449	164.494	0.1061	0.0061	77.0766	8.1762	1344.940
31	0.0521	9.4790	19.194	181.943	0.1055	0.0055	78.6395	8.2962	1509.434
32	0.0474	9.5264	21.114	201.138	0.1050	0.0050	80.1078	8.4091	1691.378
33	0.0431	9.5694	23.225	222.252	0.1045	0.0045	81.4856	8.5152	1892.515
34	0.0391	9.6086	25.548	245.477	0.1041	0.0041	82.7773	8.6149	2114.767
35	0.0356	9.6442	28.102	271.024	0.1037	0.0037	83.9872	8.7086	2360.244
36	0.0323	9.6765	30.913	299.127	0.1033	0.0033	85.1194	8.7965	2631.268
37	0.0294	9.7059	34.004	330.039	0.1030	0.0030	86.1781	8.8789	2930.395
38	0.0267	9.7327	37.404	364.043	0.1027	0.0027	87.1673	8.9562	3260.434
39	0.0243	9.7570	41.145	401.448	0.1025	0.0025	88.0908	9.0285	3624.478
40	0.0221	9.7791	45.259	442.593	0.1023	0.0023	88.9525	9.0962	4025.926
41	0.0201	9.7991	49.785	487.852	0.1020	0.0020	89.7560	9.1596	4468.518
42	0.0183	9.8174	54.764	537.637	0.1019	0.0019	90.5047	9.2188	4956.370
43	0.0166	9.8340	60.240	592.401	0.1017	0.0017	91.2019	9.2741	5494.007
44	0.0151	9.8491	66.264	652.641	0.1015	0.0015	91.8508	9.3258	6086.408
45	0.0137	9.8628	72.890	718.905	0.1014	0.0014	92.4544	9.3740	6739.048
46	0.0125	9.8753	80.180	791.795	0.1013	0.0013	93.0157	9.4190	7457.953
47	0.0113	9.8866	88.197	871.975	0.1011	0.0011	93.5372	9.4610	8249.749
48	0.0103	9.8969	97.017	960.172	0.1010	0.0010	94.0217	9.5001	9121.723
49	0.0094	9.9063	106.719	1057.190	0.1009	0.0009	94.4715	9.5365	10081.896
50	0.0085	9.9148	117.391	1163.909	0.1009	0.0009	94.8889	9.5704	11139.085
51	0.0077	9.9226	129.130	1281.299	0.1008	0.0008	95.2761	9.6020	12302.994
52	0.0070	9.9296	142.043	1410.429	0.1007	0.0007	95.6351	9.6313	13584.293
53	0.0064	9.9360	156.247	1552.472	0.1006	0.0006	95.9679	9.6586	14994.723
54	0.0058	9.9418	171.872	1708.719	0.1006	0.0006	96.2763	9.6840	16547.195

Financial Factors, Interest Rate, i = 10%

N	(P/F)	(P/A)	(F/P)	(F/A)	(A/P)	(A/F)	(P/G)	(A/G)	(F/G)
55	0.0053	9.9471	189.059	1880.591	0.1005	0.0005	96.5619	9.7075	18255.914
56	0.0048	9.9519	207.965	2069.651	0.1005	0.0005	96.8264	9.7294	20136.506
57	0.0044	9.9563	228.762	2277.616	0.1004	0.0004	97.0712	9.7497	22206.156
58	0.0040	9.9603	251.638	2506.377	0.1004	0.0004	97.2977	9.7686	24483.772
59	0.0036	9.9639	276.801	2758.015	0.1004	0.0004	97.5072	9.7861	26990.149
60	0.0033	9.9672	304.482	3034.816	0.1003	0.0003	97.7010	9.8023	29748.164
65	0.0020	9.9796	490.371	4893.707	0.1002	0.0002	98.4705	9.8672	48287.073
70	0.0013	9.9873	789.747	7887.470	0.1001	0.0001	98.9870	9.9113	78174.696
75	0.0008	9.9921	1271.895	12708.954	0.1001	0.0001	99.3317	9.9410	126339.537
80	0.0005	9.9951	2048.400	20474.002	0.1000	0.0000	99.5606	9.9609	203940.021
85	0.0003	9.9970	3298.969	32979.690	0.1000	0.0000	99.7120	9.9742	328946.903
90	0.0002	9.9981	5313.023	53120.226	0.1000	0.0000	99.8118	9.9831	530302.261
95	0.0001	9.9988	8556.676	85556.760	0.1000	0.0000	99.8773	9.9889	854617.605
100	0.0001	9.9993	13780.612	137796.123	0.1000	0.0000	99.9202	9.9927	1376961.234

Financial Factors, Interest Rate, i = 12%

N	(P/F)	(P/A)	(F/P)	(F/A)	(A/P)	(A/F)	(P/G)	(A/G)	(F/G)
1	0.8929	0.8929	1.120	1.000	1.1200	1.0000	0.0000	0.0000	0.000
2	0.7972	1.6901	1.254	2.120	0.5917	0.4717	0.7972	0.4717	1.000
3	0.7118	2.4018	1.405	3.374	0.4163	0.2963	2.2208	0.9246	3.120
4	0.6355	3.0373	1.574	4.779	0.3292	0.2092	4.1273	1.3589	6.494
5	0.5674	3.6048	1.762	6.353	0.2774	0.1574	6.3970	1.7746	11.274
6	0.5066	4.1114	1.974	8.115	0.2432	0.1232	8.9302	2.1720	17.627
7	0.4523	4.5638	2.211	10.089	0.2191	0.0991	11.6443	2.5515	25.742
8	0.4039	4.9676	2.476	12.300	0.2013	0.0813	14.4714	2.9131	35.831
9	0.3606	5.3282	2.773	14.776	0.1877	0.0677	17.3563	3.2574	48.130
10	0.3220	5.6502	3.106	17.549	0.1770	0.0570	20.2541	3.5847	62.906
11	0.2875	5.9377	3.479	20.655	0.1684	0.0484	23.1288	3.8953	80.455
12	0.2567	6.1944	3.896	24.133	0.1614	0.0414	25.9523	4.1897	101.109
13	0.2292	6.4235	4.363	28.029	0.1557	0.0357	28.7024	4.4683	125.243
14	0.2046	6.6282	4.887	32.393	0.1509	0.0309	31.3624	4.7317	153.272
15	0.1827	6.8109	5.474	37.280	0.1468	0.0268	33.9202	4.9803	185.664
16	0.1631	6.9740	6.130	42.753	0.1434	0.0234	36.3670	5.2147	222.944
17	0.1456	7.1196	6.866	48.884	0.1405	0.0205	38.6973	5.4353	265.697
18	0.1300	7.2497	7.690	55.750	0.1379	0.0179	40.9080	5.6427	314.581
19	0.1161	7.3658	8.613	63.440	0.1358	0.0158	42.9979	5.8375	370.331
20	0.1037	7.4694	9.646	72.052	0.1339	0.0139	44.9676	6.0202	433.770
21	0.0926	7.5620	10.804	81.699	0.1322	0.0122	46.8188	6.1913	505.823
22	0.0826	7.6446	12.100	92.503	0.1308	0.0108	48.5543	6.3514	587.522
23	0.0738	7.7184	13.552	104.603	0.1296	0.0096	50.1776	6.5010	680.024
24	0.0659	7.7843	15.179	118.155	0.1285	0.0085	51.6929	6.6406	784.627
25	0.0588	7.8431	17.000	133.334	0.1275	0.0075	53.1046	6.7708	902.782
26	0.0525	7.8957	19.040	150.334	0.1267	0.0067	54.4177	6.8921	1036.116
27	0.0469	7.9426	21.325	169.374	0.1259	0.0059	55.6369	7.0049	1186.450
28	0.0419	7.9844	23.884	190.699	0.1252	0.0052	56.7674	7.1098	1355.824
29	0.0374	8.0218	26.750	214.583	0.1247	0.0047	57.8141	7.2071	1546.523
30	0.0334	8.0552	29.960	241.333	0.1241	0.0041	58.7821	7.2974	1761.106
31	0.0298	8.0850	33.555	271.293	0.1237	0.0037	59.6761	7.3811	2002.438
32	0.0266	8.1116	37.582	304.848	0.1233	0.0033	60.5010	7.4586	2273.731
33	0.0238	8.1354	42.092	342.429	0.1229	0.0029	61.2612	7.5302	2578.579
34	0.0212	8.1566	47.143	384.521	0.1226	0.0026	61.9612	7.5965	2921.008

Financial Factors, Interest Rate, i = 12%

N	(P/F)	(P/A)	(F/P)	(F/A)	(A/P)	(A/F)	(P/G)	(A/G)	(F/G)
35	0.0189	8.1755	52.800	431.663	0.1223	0.0023	62.6052	7.6577	3305.529
36	0.0169	8.1924	59.136	484.463	0.1221	0.0021	63.1970	7.7141	3737.193
37	0.0151	8.2075	66.232	543.599	0.1218	0.0018	63.7406	7.7661	4221.656
38	0.0135	8.2210	74.180	609.831	0.1216	0.0016	64.2394	7.8141	4765.254
39	0.0120	8.2330	83.081	684.010	0.1215	0.0015	64.6967	7.8582	5375.085
40	0.0107	8.2438	93.051	767.091	0.1213	0.0013	65.1159	7.8988	6059.095
41	0.0096	8.2534	104.217	860.142	0.1212	0.0012	65.4997	7.9361	6826.187
42	0.0086	8.2619	116.723	964.359	0.1210	0.0010	65.8509	7.9704	7686.329
43	0.0076	8.2696	130.730	1081.083	0.1209	0.0009	66.1722	8.0019	8650.688
44	0.0068	8.2764	146.418	1211.813	0.1208	0.0008	66.4659	8.0308	9731.771
45	0.0061	8.2825	163.988	1358.230	0.1207	0.0007	66.7342	8.0572	10943.584
46	0.0054	8.2880	183.666	1522.218	0.1207	0.0007	66.9792	8.0815	12301.814
47	0.0049	8.2928	205.706	1705.884	0.1206	0.0006	67.2028	8.1037	13824.031
48	0.0043	8.2972	230.391	1911.590	0.1205	0.0005	67.4068	8.1241	15529.915
49	0.0039	8.3010	258.038	2141.981	0.1205	0.0005	67.5929	8.1427	17441.505
50	0.0035	8.3045	289.002	2400.018	0.1204	0.0004	67.7624	8.1597	19583.485
51	0.0031	8.3076	323.682	2689.020	0.1204	0.0004	67.9169	8.1753	21983.504
52	0.0028	8.3103	362.524	3012.703	0.1203	0.0003	68.0576	8.1895	24672.524
53	0.0025	8.3128	406.027	3375.227	0.1203	0.0003	68.1856	8.2025	27685.227
54	0.0022	8.3150	454.751	3781.255	0.1203	0.0003	68.3022	8.2143	31060.454
55	0.0020	8.3170	509.321	4236.005	0.1202	0.0002	68.4082	8.2251	34841.709
56	0.0018	8.3187	570.439	4745.326	0.1202	0.0002	68.5046	8.2350	39077.714
57	0.0016	8.3203	638.892	5315.765	0.1202	0.0002	68.5923	8.2440	43823.039
58	0.0014	8.3217	715.559	5954.656	0.1202	0.0002	68.6719	8.2522	49138.804
59	0.0012	8.3229	801.426	6670.215	0.1201	0.0001	68.7443	8.2596	55093.461
60	0.0011	8.3240	897.597	7471.641	0.1201	0.0001	68.8100	8.2664	61763.676
65	0.0006	8.3281	1581.872	13173.937	0.1201	0.0001	69.0581	8.2922	109241.145
70	0.0004	8.3303	2787.800	23223.332	0.1200	0.0000	69.2103	8.3082	192944.432
75	0.0002	8.3316	4913.056	40933.799	0.1200	0.0000	69.3031	8.3181	340489.989
80	0.0001	8.3324	8658.483	72145.693	0.1200	0.0000	69.3594	8.3241	600547.438
85	0.0001	8.3328	15259.206	127151.714	0.1200	0.0000	69.3935	8.3278	1058889.283
90	0.0000	8.3330	26891.934	224091.119	0.1200	0.0000	69.4140	8.3300	1866675.988
95	0.0000	8.3332	47392.777	394931.472	0.1200	0.0000	69.4263	8.3313	3290303.932
100	0.0000	8.3332	83522.266	696010.548	0.1200	0.0000	69.4336	8.3321	5799254.564

Financial Factors, Interest Rate, i = 15%

N	(P/F)	(P/A)	(F/P)	(F/A)	(A/P)	(A/F)	(P/G)	(A/G)	(F/G)
1	0.8696	0.8696	1.150	1.000	1.1500	1.0000	0.0000	0.0000	0.00
2	0.7561	1.6257	1.323	2.150	0.6151	0.4651	0.7561	0.4651	1.00
3	0.6575	2.2832	1.521	3.473	0.4380	0.2880	2.0712	0.9071	3.15
4	0.5718	2.8550	1.749	4.993	0.3503	0.2003	3.7864	1.3263	6.62
5	0.4972	3.3522	2.011	6.742	0.2983	0.1483	5.7751	1.7228	11.62
6	0.4323	3.7845	2.313	8.754	0.2642	0.1142	7.9368	2.0972	18.36
7	0.3759	4.1604	2.660	11.067	0.2404	0.0904	10.1924	2.4498	27.11
8	0.3269	4.4873	3.059	13.727	0.2229	0.0729	12.4807	2.7813	38.18
9	0.2843	4.7716	3.518	16.786	0.2096	0.0596	14.7548	3.0922	51.91
10	0.2472	5.0188	4.046	20.304	0.1993	0.0493	16.9795	3.3832	68.69
11	0.2149	5.2337	4.652	24.349	0.1911	0.0411	19.1289	3.6549	89.00
12	0.1869	5.4206	5.350	29.002	0.1845	0.0345	21.1849	3.9082	113.34
13	0.1625	5.5831	6.153	34.352	0.1791	0.0291	23.1352	4.1438	142.35
14	0.1413	5.7245	7.076	40.505	0.1747	0.0247	24.9725	4.3624	176.70

Financial Factors, Interest Rate, i = 15%

N	(P/F)	(P/A)	(F/P)	(F/A)	(A/P)	(A/F)	(P/G)	(A/G)	(F/G)
15	0.1229	5.8474	8.137	47.580	0.1710	0.0210	26.6930	4.5650	217.20
16	0.1069	5.9542	9.358	55.717	0.1679	0.0179	28.2960	4.7522	264.78
17	0.0929	6.0472	10.761	65.075	0.1654	0.0154	29.7828	4.9251	320.50
18	0.0808	6.1280	12.375	75.836	0.1632	0.0132	31.1565	5.0843	385.58
19	0.0703	6.1982	14.232	88.212	0.1613	0.0113	32.4213	5.2307	461.41
20	0.0611	6.2593	16.367	102.444	0.1598	0.0098	33.5822	5.3651	549.62
21	0.0531	6.3125	18.822	118.810	0.1584	0.0084	34.6448	5.4883	652.07
22	0.0462	6.3587	21.645	137.632	0.1573	0.0073	35.6150	5.6010	770.88
23	0.0402	6.3988	24.891	159.276	0.1563	0.0063	36.4988	5.7040	908.51
24	0.0349	6.4338	28.625	184.168	0.1554	0.0054	37.3023	5.7979	1067.79
25	0.0304	6.4641	32.919	212.793	0.1547	0.0047	38.0314	5.8834	1251.95
26	0.0264	6.4906	37.857	245.712	0.1541	0.0041	38.6918	5.9612	1464.75
27	0.0230	6.5135	43.535	283.569	0.1535	0.0035	39.2890	6.0319	1710.46
28	0.0200	6.5335	50.066	327.104	0.1531	0.0031	39.8283	6.0960	1994.03
29	0.0174	6.5509	57.575	377.170	0.1527	0.0027	40.3146	6.1541	2321.13
30	0.0151	6.5660	66.212	434.745	0.1523	0.0023	40.7526	6.2066	2698.30
31	0.0131	6.5791	76.144	500.957	0.1520	0.0020	41.1466	6.2541	3133.05
32	0.0114	6.5905	87.565	577.100	0.1517	0.0017	41.5006	6.2970	3634.00
33	0.0099	6.6005	100.700	664.666	0.1515	0.0015	41.8184	6.3357	4211.10
34	0.0086	6.6091	115.805	765.365	0.1513	0.0013	42.1033	6.3705	4875.77
35	0.0075	6.6166	133.176	881.170	0.1511	0.0011	42.3586	6.4019	5641.13
36	0.0065	6.6231	153.152	1014.346	0.1510	0.0010	42.5872	6.4301	6522.30
37	0.0057	6.6288	176.125	1167.498	0.1509	0.0009	42.7916	6.4554	7536.65
38	0.0049	6.6338	202.543	1343.622	0.1507	0.0007	42.9743	6.4781	8704.15
39	0.0043	6.6380	232.925	1546.165	0.1506	0.0006	43.1374	6.4985	10047.77
40	0.0037	6.6418	267.864	1779.090	0.1506	0.0006	43.2830	6.5168	11593.94
41	0.0032	6.6450	308.043	2046.954	0.1505	0.0005	43.4128	6.5331	13373.03
42	0.0028	6.6478	354.250	2354.997	0.1504	0.0004	43.5286	6.5478	15419.98
43	0.0025	6.6503	407.387	2709.246	0.1504	0.0004	43.6317	6.5609	17774.98
44	0.0021	6.6524	468.495	3116.633	0.1503	0.0003	43.7235	6.5725	20484.22
45	0.0019	6.6543	538.769	3585.128	0.1503	0.0003	43.8051	6.5830	23600.86
46	0.0016	6.6559	619.585	4123.898	0.1502	0.0002	43.8778	6.5923	27185.98
47	0.0014	6.6573	712.522	4743.482	0.1502	0.0002	43.9423	6.6006	31309.88
48	0.0012	6.6585	819.401	5456.005	0.1502	0.0002	43.9997	6.6080	36053.36
49	0.0011	6.6596	942.311	6275.405	0.1502	0.0002	44.0506	6.6146	41509.37
50	0.0009	6.6605	1083.657	7217.716	0.1501	0.0001	44.0958	6.6205	47784.78
51	0.0008	6.6613	1246.206	8301.374	0.1501	0.0001	44.1360	6.6257	55002.49
52	0.0007	6.6620	1433.137	9547.580	0.1501	0.0001	44.1715	6.6304	63303.87
53	0.0006	6.6626	1648.108	10980.717	0.1501	0.0001	44.2031	6.6345	72851.44
54	0.0005	6.6631	1895.324	12628.824	0.1501	0.0001	44.2311	6.6382	83832.16
55	0.0005	6.6636	2179.622	14524.148	0.1501	0.0001	44.2558	6.6414	96460.99
56	0.0004	6.6640	2506.566	16703.770	0.1501	0.0001	44.2778	6.6443	110985.13
57	0.0003	6.6644	2882.550	19210.336	0.1501	0.0001	44.2972	6.6469	127688.90
58	0.0003	6.6647	3314.933	22092.886	0.1500	0.0000	44.3144	6.6492	146899.24
59	0.0003	6.6649	3812.173	25407.819	0.1500	0.0000	44.3296	6.6512	168992.13
60	0.0002	6.6651	4383.999	29219.992	0.1500	0.0000	44.3431	6.6530	194399.94
65	0.0001	6.6659	8817.787	58778.583	0.1500	0.0000	44.3903	6.6593	391423.88
70	0.0001	6.6663	17735.720	118231.467	0.1500	0.0000	44.4156	6.6627	787743.11
75	0.0000	6.6665	35672.868	237812.453	0.1500	0.0000	44.4292	6.6646	1584916.35
80	0.0000	6.6666	71750.879	478332.529	0.1500	0.0000	44.4364	6.6656	3188350.20
85	0.0000	6.6666	144316.647	962104.313	0.1500	0.0000	44.4402	6.6661	6413462.09
90	0.0000	6.6666	290272.325	1935142.168	0.1500	0.0000	44.4422	6.6664	12900347.79
95	0.0000	6.6667	583841.328	3892268.851	0.1500	0.0000	44.4433	6.6665	25947825.67
100	0.0000	6.6667	1174313.451	7828749.671	0.1500	0.0000	44.4438	6.6666	52190997.81

Financial Factors, Interest Rate, i = 20%

N	(P/F)	(P/A)	(F/P)	(F/A)	(A/P)	(A/F)	(P/G)	(A/G)	(F/G)
1	0.8333	0.8333	1.200	1.000	1.2000	1.0000	0.0000	0.0000	0.0000
2	0.6944	1.5278	1.440	2.200	0.6545	0.4545	0.6944	0.4545	1.0000
3	0.5787	2.1065	1.728	3.640	0.4747	0.2747	1.8519	0.8791	3.2000
4	0.4823	2.5887	2.074	5.368	0.3863	0.1863	3.2986	1.2742	6.8400
5	0.4019	2.9906	2.488	7.442	0.3344	0.1344	4.9061	1.6405	12.2080
6	0.3349	3.3255	2.986	9.930	0.3007	0.1007	6.5806	1.9788	19.6496
7	0.2791	3.6046	3.583	12.916	0.2774	0.0774	8.2551	2.2902	29.5795
8	0.2326	3.8372	4.300	16.499	0.2606	0.0606	9.8831	2.5756	42.4954
9	0.1938	4.0310	5.160	20.799	0.2481	0.0481	11.4335	2.8364	58.9945
10	0.1615	4.1925	6.192	25.959	0.2385	0.0385	12.8871	3.0739	79.7934
11	0.1346	4.3271	7.430	32.150	0.2311	0.0311	14.2330	3.2893	105.7521
12	0.1122	4.4392	8.916	39.581	0.2253	0.0253	15.4667	3.4841	137.9025
13	0.0935	4.5327	10.699	48.497	0.2206	0.0206	16.5883	3.6597	177.48
14	0.0779	4.6106	12.839	59.196	0.2169	0.0169	17.6008	3.8175	225.98
15	0.0649	4.6755	15.407	72.035	0.2139	0.0139	18.5095	3.9588	285.18
16	0.0541	4.7296	18.488	87.442	0.2114	0.0114	19.3208	4.0851	357.21
17	0.0451	4.7746	22.186	105.931	0.2094	0.0094	20.0419	4.1976	444.65
18	0.0376	4.8122	26.623	128.117	0.2078	0.0078	20.6805	4.2975	550.58
19	0.0313	4.8435	31.948	154.740	0.2065	0.0065	21.2439	4.3861	678.70
20	0.0261	4.8696	38.338	186.688	0.2054	0.0054	21.7395	4.4643	833.44
21	0.0217	4.8913	46.005	225.026	0.2044	0.0044	22.1742	4.5334	1020.13
22	0.0181	4.9094	55.206	271.031	0.2037	0.0037	22.5546	4.5941	1245.15
23	0.0151	4.9245	66.247	326.237	0.2031	0.0031	22.8867	4.6475	1516.18
24	0.0126	4.9371	79.497	392.484	0.2025	0.0025	23.1760	4.6943	1842.42
25	0.0105	4.9476	95.396	471.981	0.2021	0.0021	23.4276	4.7352	2234.91
26	0.0087	4.9563	114.475	567.377	0.2018	0.0018	23.6460	4.7709	2706.89
27	0.0073	4.9636	137.371	681.853	0.2015	0.0015	23.8353	4.8020	3274.26
28	0.0061	4.9697	164.845	819.223	0.2012	0.0012	23.9991	4.8291	3956.12
29	0.0051	4.9747	197.814	984.068	0.2010	0.0010	24.1406	4.8527	4775.34
30	0.0042	4.9789	237.376	1181.882	0.2008	0.0008	24.2628	4.8731	5759.41
31	0.0035	4.9824	284.852	1419.258	0.2007	0.0007	24.3681	4.8908	6941.29
32	0.0029	4.9854	341.822	1704.109	0.2006	0.0006	24.4588	4.9061	8360.55
33	0.0024	4.9878	410.186	2045.931	0.2005	0.0005	24.5368	4.9194	10064.66
34	0.0020	4.9898	492.224	2456.118	0.2004	0.0004	24.6038	4.9308	12110.59
35	0.0017	4.9915	590.668	2948.341	0.2003	0.0003	24.6614	4.9406	14566.71
36	0.0014	4.9929	708.802	3539.009	0.2003	0.0003	24.7108	4.9491	17515.05
37	0.0012	4.9941	850.562	4247.811	0.2002	0.0002	24.7531	4.9564	21054.06
38	0.0010	4.9951	1020.675	5098.373	0.2002	0.0002	24.7894	4.9627	25301.87
39	0.0008	4.9959	1224.810	6119.048	0.2002	0.0002	24.8204	4.9681	30400.24
40	0.0007	4.9966	1469.772	7343.858	0.2001	0.0001	24.8469	4.9728	36519.29
41	0.0006	4.9972	1763.726	8813.629	0.2001	0.0001	24.8696	4.9767	43863.15
42	0.0005	4.9976	2116.471	10577.355	0.2001	0.0001	24.8890	4.9801	52676.78
43	0.0004	4.9980	2539.765	12693.826	0.2001	0.0001	24.9055	4.9831	63254.13
44	0.0003	4.9984	3047.718	15233.592	0.2001	0.0001	24.9196	4.9856	75947.96
45	0.0003	4.9986	3657.262	18281.310	0.2001	0.0001	24.9316	4.9877	91181.55
46	0.0002	4.9989	4388.714	21938.572	0.2000	0.0000	24.9419	4.9895	109462.86
47	0.0002	4.9991	5266.457	26327.286	0.2000	0.0000	24.9506	4.9911	131401.43
48	0.0002	4.9992	6319.749	31593.744	0.2000	0.0000	24.9581	4.9924	157728.72
49	0.0001	4.9993	7583.698	37913.492	0.2000	0.0000	24.9644	4.9935	189322.46
50	0.0001	4.9995	9100.438	45497.191	0.2000	0.0000	24.9698	4.9945	227235.95
51	0.0001	4.9995	10920.526	54597.629	0.2000	0.0000	24.9744	4.9953	272733.14
52	0.0001	4.9996	13104.631	65518.155	0.2000	0.0000	24.9783	4.9960	327330.77
53	0.0001	4.9997	15725.557	78622.786	0.2000	0.0000	24.9816	4.9966	392848.93
54	0.0001	4.9997	18870.669	94348.343	0.2000	0.0000	24.9844	4.9971	471471.71

Financial Factors, Interest Rate, i = 20%

N	(P/F)	(P/A)	(F/P)	(F/A)	(A/P)	(A/F)	(P/G)	(A/G)	(F/G)
55	0.0000	4.9998	22644.802	113219.011	0.2000	0.0000	24.9868	4.9976	565820.06
56	0.0000	4.9998	27173.763	135863.814	0.2000	0.0000	24.9888	4.9979	679039.07
57	0.0000	4.9998	32608.515	163037.576	0.2000	0.0000	24.9905	4.9983	814902.88
58	0.0000	4.9999	39130.218	195646.092	0.2000	0.0000	24.9919	4.9985	977940.46
59	0.0000	4.9999	46956.262	234776.310	0.2000	0.0000	24.9932	4.9987	1173586.55
60	0.0000	4.9999	56347.514	281732.572	0.2000	0.0000	24.9942	4.9989	1408362.86
65	0.0000	5.0000	140210.647	701048.235	0.2000	0.0000	24.9975	4.9995	3504916.17
70	0.0000	5.0000	348888.957	1744439.785	0.2000	0.0000	24.9989	4.9998	8721848.92
75	0.0000	5.0000	868147.369	4340731.847	0.2000	0.0000	24.9995	4.9999	21703284.23
80	0.0000	5.0000	2160228.462	10801137.310	0.2000	0.0000	24.9998	5.0000	54005286.55
85	0.0000	5.0000	5375339.687	26876693.433	0.2000	0.0000	24.9999	5.0000	134383042.16
90	0.0000	5.0000	13375565.249	66877821.245	0.2000	0.0000	25.0000	5.0000	334388656.22
95	0.0000	5.0000	33282686.520	166413427.601	0.2000	0.0000	25.0000	5.0000	832066663.01
100	0.0000	5.0000	82817974.522	414089867.610	0.2000	0.0000	25.0000	5.0000	2070448838.05

Financial Factors for Interest Rate, i = 25%

N	(P/F)	(P/A)	(F/P)	(F/A)	(A/P)	(A/F)	(P/G)	(A/G)	(F/G)
1	0.8000	0.8000	1.25	1.00	1.250	1.000	0.000	0.0000	0.00
2	0.6400	1.4400	1.56	2.25	0.694	0.444	0.640	0.4444	1.00
3	0.5120	1.9520	1.95	3.81	0.512	0.262	1.664	0.8525	3.25
4	0.4096	2.3616	2.44	5.77	0.423	0.173	2.893	1.2249	7.06
5	0.3277	2.6893	3.05	8.21	0.372	0.122	4.204	1.5631	12.83
6	0.2621	2.9514	3.81	11.26	0.339	0.089	5.514	1.8683	21.04
7	0.2097	3.1611	4.77	15.07	0.316	0.066	6.773	2.1424	32.29
8	0.1678	3.3289	5.96	19.84	0.300	0.050	7.947	2.3872	47.37
9	0.1342	3.4631	7.45	25.80	0.289	0.039	9.021	2.6048	67.21
10	0.1074	3.5705	9.31	33.25	0.280	0.030	9.987	2.7971	93.01
11	0.0859	3.6564	11.64	42.57	0.273	0.023	10.846	2.9663	126.26
12	0.0687	3.7251	14.55	54.21	0.268	0.018	11.602	3.1145	168.83
13	0.0550	3.7801	18.19	68.76	0.265	0.015	12.262	3.2437	223.04
14	0.0440	3.8241	22.74	86.95	0.262	0.012	12.833	3.3559	291.80
15	0.0352	3.8593	28.42	109.69	0.259	0.009	13.326	3.4530	378.75
16	0.0281	3.8874	35.53	138.11	0.257	0.007	13.748	3.5366	488.43
17	0.0225	3.9099	44.41	173.64	0.256	0.006	14.108	3.6084	626.54
18	0.0180	3.9279	55.51	218.04	0.255	0.005	14.415	3.6698	800.18
19	0.0144	3.9424	69.39	273.56	0.254	0.004	14.674	3.7222	1018.22
20	0.0115	3.9539	86.74	342.94	0.253	0.003	14.893	3.7667	1291.78
21	0.0092	3.9631	108.42	429.68	0.252	0.002	15.078	3.8045	1634.72
22	0.0074	3.9705	135.53	538.10	0.252	0.002	15.233	3.8365	2064.40
23	0.0059	3.9764	169.41	673.63	0.251	0.001	15.362	3.8634	2602.51
24	0.0047	3.9811	211.76	843.03	0.251	0.001	15.471	3.8861	3276.13
25	0.0038	3.9849	264.70	1054.79	0.251	0.001	15.562	3.9052	4119.16
26	0.0030	3.9879	330.87	1319.49	0.251	0.001	15.637	3.9212	5173.96
27	0.0024	3.9903	413.59	1650.36	0.251	0.001	15.700	3.9346	6493.44
28	0.0019	3.9923	516.99	2063.95	0.250	0.000	15.752	3.9457	8143.81
29	0.0015	3.9938	646.23	2580.94	0.250	0.000	15.796	3.9551	10207.76
30	0.0012	3.9950	807.79	3227.17	0.250	0.000	15.832	3.9628	12788.70
31	0.0010	3.9960	1009.74	4034.97	0.250	0.000	15.861	3.9693	16015.87
32	0.0008	3.9968	1262.18	5044.71	0.250	0.000	15.886	3.9746	20050.84
33	0.0006	3.9975	1577.72	6306.89	0.250	0.000	15.906	3.9791	25095.55

Financial Factors for Interest Rate, i = 25%

N	(P/F)	(P/A)	(F/P)	(F/A)	(A/P)	(A/F)	(P/G)	(A/G)	(F/G)
34	0.0005	3.9980	1972.15	7884.61	0.250	0.000	15.923	3.9828	31402.44
35	0.0004	3.9984	2465.19	9856.76	0.250	0.000	15.937	3.9858	39287.05
36	0.0003	3.9987	3081.49	12321.95	0.250	0.000	15.948	3.9883	49143.81
37	0.0003	3.9990	3851.86	15403.44	0.250	0.000	15.957	3.9904	61465.76
38	0.0002	3.9992	4814.82	19255.30	0.250	0.000	15.965	3.9921	76869.20
39	0.0002	3.9993	6018.53	24070.12	0.250	0.000	15.971	3.9935	96124.50
40	0.0001	3.9995	7523.16	30088.66	0.250	0.000	15.977	3.9947	120194.62
41	0.0001	3.9996	9403.95	37611.82	0.250	0.000	15.981	3.9956	150283.28
42	0.0001	3.9997	11754.94	47015.77	0.250	0.000	15.984	3.9964	187895.10
43	0.0001	3.9997	14693.68	58770.72	0.250	0.000	15.987	3.9971	234910.87
44	0.0001	3.9998	18367.10	73464.40	0.250	0.000	15.990	3.9976	293681.59
45	0.0000	3.9998	22958.87	91831.50	0.250	0.000	15.991	3.9980	367145.98
46	0.0000	3.9999	28698.59	114790.37	0.250	0.000	15.993	3.9984	458977.48
47	0.0000	3.9999	35873.24	143488.96	0.250	0.000	15.994	3.9987	573767.85
48	0.0000	3.9999	44841.55	179362.20	0.250	0.000	15.995	3.9989	717256.81
49	0.0000	3.9999	56051.94	224203.75	0.250	0.000	15.996	3.9991	896619.02
50	0.0000	3.9999	70064.92	280255.69	0.250	0.000	15.997	3.9993	1120822.77
51	0.0000	4.0000	87581.15	350320.62	0.250	0.000	15.997	3.9994	1401078.46
52	0.0000	4.0000	109476.44	437901.77	0.250	0.000	15.998	3.9995	1751399.08
53	0.0000	4.0000	136845.55	547378.21	0.250	0.000	15.998	3.9996	2189300.85
54	0.0000	4.0000	171056.94	684223.77	0.250	0.000	15.999	3.9997	2736679.06
55	0.0000	4.0000	213821.18	855280.71	0.250	0.000	15.999	3.9997	3420902.83
56	0.0000	4.0000	267276.47	1069101.88	0.250	0.000	15.999	3.9998	4276183.54
57	0.0000	4.0000	334095.59	1336378.36	0.250	0.000	15.999	3.9998	5345285.42
58	0.0000	4.0000	417619.49	1670473.94	0.250	0.000	15.999	3.9999	6681663.78
59	0.0000	4.0000	522024.36	2088093.43	0.250	0.000	16.000	3.9999	8352137.72
60	0.0000	4.0000	652530.45	2610117.79	0.250	0.000	16.000	3.9999	10440231.15
65	0.0000	4.0000	1991364.89	7965455.56	0.250	0.000	16.000	4.0000	31861562.22
70	0.0000	4.0000	6077163.36	24308649.43	0.250	0.000	16.000	4.0000	97234317.72
75	0.0000	4.0000	18546030.75	74184119.01	0.250	0.000	16.000	4.0000	296736176.05
80	0.0000	4.0000	56597994.24	226391972.97	0.250	0.000	16.000	4.0000	905567571.88
85	0.0000	4.0000	172723371.10	690893480.41	0.250	0.000	16.000	4.0000	2763573581.63
90	0.0000	4.0000	527109897.16	2108439584.65	0.250	0.000	16.000	4.0000	8433757978.58
95	0.0000	4.0000	1608611746.71	6434446982.84	0.250	0.000	16.000	4.0000	25737787551.34
100	0.0000	4.0000	4909093465.30	19636373857.19	0.250	0.000	16.000	4.0000	78545495028.76

Index

R